Handbook of Current Affairs and General Agriculture

Handbook of Current Affairs and General Agriculture

Gurdeep Lal
Agricultural Field Officer
Bank of Baroda, Lakhimpur Kheri (U.P.)

Rashmi Sahu
SRF at CIAE Bhopal (M.P.)

Shivendu Pratap Singh Solanki, Ph.D.
Scholar, Punjab Agricultural University, Ludhiana (Punjab)

SSPH

SATISH SERIAL PUBLISHING HOUSE
403, Express Tower, Commercial Complex
Azadpur, Delhi-110033 (India)
Phone : 011-27672852, Fax : 91-11-27672046
e-mail : info@satishserial.com, hkjain1975@yahoo.com
Website : www.satishserial.com

Published by :

SATISH SERIAL PUBLISHING HOUSE
403, Express Tower, Commercial Complex, Azadpur, Delhi-110033 (INDIA)
Phone : 011-27672852 Fax : 91-11-27672046
E-mail : info@satishserial.com, hkjain1975@yahoo.com

ISBN 978-93-88020-33-6

Composed, Designed & Printed in India

Preface

This is the epoch of smart and tough competition. Competition is exploding in all fronts and in the modern times everyone wants to get succeed in competition. Agriculture is the vast subject including Agricultural Current Affairs and General Agriculture for preparation of competitive examination. To get the success in agricultural competition becoming selective for study is the astute move and the only conduit for success. The ICAR, New Delhi is conducting various examinations like ICAR-JRF, SRF, NET examination, etc. In addition to that Government of India and State Government conduct various exams like IFS, FCI, AIC, IFFCO, NSCI, ADO, D.H.O, AEO and all Similar Competitive Agricultural Related Exams and IBPS-AFO Exam every year conducted by Government of India. All these examinations are mostly based on Agricultural Current Affairs and General Agriculture questions and students always look for study material that is ready to read and directly useful for examination. Keeping in view, it felt to bring out a book which could provide the basic and innovative knowledge of various fields of agriculture and allied sciences. After success of our previous book "**Question Bank on General Agriculture and Engineering**" that was prepared in most simple and appropriate manner by taking the view of students, teachers and scientists. In the same manner, we are bringing a new book "**Handbook of Current Affairs and General Agriculture**". We hope that the book will fulfill the need of students appearing in agricultural based competitive examinations. The help and prop up extended by our family members, friends, student, teachers and scientists during the compilation of this book is highly appreciable. We humbly welcome the valuable suggestions from readers for further improvement of this book.

About the Authors

Gurdeep Lal, working as an Agricultural Field Officer, Bank of Baroda, born at village Kazibans, Saharanpur District, Uttar Pradesh. He received his Post Graduation degree (2015-17) in Farm Machinery and Power Engineering from Department of Agriculture and Food Engineering IIT, Kharagpur, West Bengal and Under Graduate degree (2011-15) in Agricultural Engineering from College of Agricultural Engineering and Technology, Vasantrao Naik Marathwada Agricultural University, Parbhani (Maharashtra). He was awarded with ICAR-UG-NTS for his B.Tech. His schooling done in Jawahar Navodaya Vidyalaya, Saharanpur (2004-2011). He has qualified the GATE, ICAR - JRF, in the year 2015. He is the founder and tutor of YouTube Channel Advance Agricultural Classes, also mentor of *http//www.advanceagriclasses.com*.

Rashmi Sahu, working as an SRF at CIAE Bhopal, MP, born at District Jabalpur, Madhya Pradesh. She received her Post Graduation degree (2015-17) in Farm Machinery and Power Engineering from Department of Agriculture and Food Engineering IIT, Kharagpur, West Bengal and Under Graduate degree (2011-15) in Agricultural Engineering from College of Agricultural Engineering, Jawaharlal Nehru Krishi Vishwa Vidyalaya, Jabalpur (Madhya Pradesh). She has qualified the GATE AIR 23, ICAR-JRF AIR 70, in the year 2015.

Shivendu Pratap Singh Solanki, Ph.D. Scholar, Punjab Agricultural University, Ludhiana, born at village Lamkana, Katni District, Madhya Pradesh. He received his under graduate degree in Horticulture (2011-15) from College of Horticulture, Vasantrao Naik Marathwada Agricultural University, Parbhani (Maharashtra) and awarded with University Gold Medal and Best Student award for excellent academic performance. He was awarded the ICAR-JRF for his M.Sc. (Horticulture) in Fruit Science at Dr. Yashwant Singh Parmar University of Horticulture and Forestry, Nauni, Solan (Himachal Pradesh) in 2015-17. He has qualified the ICAR-ASRB-NET in Fruit Science and also qualified ICAR-PGS-SRF examination with AIR 2nd rank in the year 2017. He is advisory board member and reviewer of International Journal of Agricultural Sciences and Assistant Editor of International Journal of Medicinal Plant Studies. He is also a life member of various reputed scientific societies. He has published 11 popular articles and 8 research articles in national and international journals. And also awarded with Excellence AIASA, Cabinet Member Award-2017.

Contents

Chapter 1

Union Budget Highlights of Agriculture and Rural Economy

- The Union Minister for Finance and Corporate Affairs Shri Arun Jaitley while presenting the General Budget, 2018-19 in Parliament said that Indian society, polity and economy had shown remarkable resilience in adjusting with the structural reforms.
- IMF, in its latest Update, has forecast that India will grow at 7.4% next year in the backdrop of services resuming high growth rates of 8% plus, exports expected to grow at 15% in 2017-18 and manufacturing back on good growth path.
- Referring to the Government's commitment to the welfare of farmers and doubling farmers' income by 2022, the Finance Minister announced a slew of new schemes and measures.
- Government has decided to keep MSP for all unannounced Kharif crops atleast one and half times of their production cost after declaring the same for the majority of Rabi Crops.
- The volume of institutional credit for agriculture sector from year-to-year increased from Rs.8.5 lakh crore in 2014-15 to Rs.10 lakh crore in 2017-18 and he proposed to raise this to Rs.11 lakh crore for the year 2018-19.
- After the establishment of Dairy Infrastructure Fund, setting up a Fisheries and Aquaculture Infrastructure Development Fund (FAIDF) for fisheries sector and an Animal Husbandry Infrastructure Development Fund (AHIDF) for financing infrastructure requirement of animal husbandry sector with a total corpus of Rs.10,000 crore for the two new funds.
- On the lines of "Operation Flood" a new Scheme "Operation Greens" was announced with an outlay of Rs. 500 Crore to address the challenge of price volatility of perishable commodities like tomato, onion and potato with the satisfaction of both the farmers and consumers.

- To develop and upgrade existing 22,000 rural haats into Gramin Agricultural Markets (grams) to take care of the interests of more than 86% small and marginal farmers.
- These grams, electronically linked to e-NAM and exempted from regulations of apmcs, will provide farmers facility to make direct sale to consumers and bulk purchasers.
- An Agri-Market Infrastructure Fund with a corpus of Rs.2000 crore will be setup for developing and upgrading agricultural marketing infrastructure in the 22,000 Gramin Agricultural Markets (grams) and 585 apmcs. He said, so far 470 apmcs have been connected to e-NAM network and rest will be connected by March, 2018.
- Rs. 200 crore for organized cultivation of highly specialized medicinal and aromatic plants and said that the organic farming by Farmer Producer Organizations (VPOs) and Village Producers' Organizations (VPOs) in large clusters, preferably of 1000 hectares each will be encouraged.
- Allocation for Ministry of Food Processing has been doubled from Rs.715 crore in 2017-18 to Rs.1400 crore in 2018-19.
- Terming Bamboo as 'Green Gold', the Finance Minister announced a Re-structured National Bamboo Mission with an outlay of Rs.1290 crore to promote bamboo sector in a holistic manner.
- Under Prime Minister Krishi Sinchai Yojna- *Har Khet ko Pani*, 96 deprived irrigation districts will be taken up with an allocation of Rs. 2600 crore.
- The Centre will work with the state governments to facilitate farmers for installing solar water pumps to irrigate their fields. He also proposed to extend the facility of Kisan Credit Cards to fisheries and animal husbandry farmers to help them meet their working capital needs.
- India's agri-exports potential is as high as US $100 billion against current exports of US $30 billion and to realize this potential, export of agri-commodities will be liberalized.
- It is proposed to set up state-of-the-art testing facilities in all the 42 Mega Food Parks.

- A special Scheme to support the efforts of the governments of Haryana, Punjab, Uttar Pradesh and the NCT of Delhi to address air pollution in the Delhi-NCR region by subsidizing machinery required for in-situ management of crop residue.
- On the loans to Self Help Groups of women, the Finance Minister said it increased to about Rupees 42,500 crore in 2016-17, growing 37% over previous year and expressed confidence that loans to shgs will increase to Rs. 75,000 crore by March, 2019.
- There is a substantially increased allocation of National Rural Livelihood Mission to Rs. 5750 crore in 2018-19.
- Referring to the measures taken for the benefit of lower and middle class, the Finance Minister said, under Ujjwala Scheme distribution of free LPG connections will be given to 8 crore poor women instead of the previous target of 5 crore women.
- Under Saubhagya Yojna, 4 crore poor households are being provided with electricity connection with an outlay of Rs.16,000 crore. To fulfill target of housing for all by 2022, more than one crore houses will be built by 2019 in rural areas, besides already constructed 6 crore toilets under Swachh Bharat Mission.
- Next year Government will be on providing maximum livelihood opportunities in the rural areas by spending more on livelihood, agriculture and allied activities and construction of rural infrastructure. In the year 2018 19, for creation of livelihood and infrastructure in rural areas, total amount to be spent by the Ministries will be Rs.14.34 lakh crore, including extra-budgetary and non-budgetary resources of Rs.11.98 lakh crore.
- Apart from employment due to farming activities and self-employment, this expenditure will create employment of 321 crore person days, 3.17 lakh kilometres of rural roads, 51 lakh new rural houses, 1.88 crore toilets, and provide 1.75 crore new household electric connections besides boosting agricultural growth.

❏❏❏

Chapter 2

World Agricultural Scenario and Position of India

Important Facts of Agriculture

1. India stands 100th position in Global hunger index (2017)
2. GDP Growth of India in 2016-17 was 7.1 % and present is 6.7% (2017-18)
3. GVA contribution of Agriculture & Allied sectors-17.5% (2015-16) and 17.4 % (2016-17)
4. GVA growth rate of Agriculture and allied sector-0.7% (2015-16) and 4.9 % (2016-17)
5. Total production of milk: 163.7.5 mt (2015-16)
6. Per capita availability of milk: 352 gm/day
7. Total production of coffee in 2017-18 was 316,000 metric tonnes (MT)
8. India stands 7th in coffee production and first is Brazil
9. Karnataka stands first in coffee production in India
10. Total egg production – 88.1 billion
11. Per capita availability of egg - 69 (number/annum)
12. Tamil Nadu stands first in egg production
13. India stands 3rd in Egg production
14. Total fish production in 11.41mt (2016-17) i.e. around 6.4 % of world fish production
15. Marine fish production in the country stood at 3.83 mt in 2015-16
16. Total wool production in 2016-17 is 43.50 million kg (raw wool)

17. Top wool-producing state in India is Rajasthan
18. India is the 7th largest producer of wool and contributes 1.8 per cent to total world production of wool
19. Production of natural rubber in 2017-18 was 8 lakh tonnes
20. Rubber Productivity during 2017-18- 1449 kg/ha
21. World Ranking in Natural Rubber Production – Sixth
22. World Ranking in Natural Rubber Consumption - Second
23. Total meat production in 2016-17 is 7.38 million tons (Target estimated)
24. Uttar Pradesh is top meat producing state in India
25. Highest exported: Cereals
26. Highest imported: Edible oil and pulses
27. There are 46 export authorities in India
28. Cropping intensity of India is 142 % (2016-17)
29. The government in the Budget had enhanced the corpus of the Rural Infrastructure Development Fund (RIFD) to Rs. 280 billion with an overall objective of doubling farmers income by 2022.
30. NABARD is the main agency for implementing RIFD scheme.
31. Harsh Kumar Bhanwala is the Chairman of NABARD. Its headquarter is in Mumbai, Maharashtra.
32. UNEP was founded as result of United Nations Conference on Human Environment in June 1972.
33. It is headquartered in Nairobi, Kenya.
34. Radha Mohan Singh, Hon'ble Minister of Agriculture.
35. President of the World Bank- Jim Yong Kim
36. Headquarters of World Bank: Washington, DC
37. The Ministry of Agriculture and Farmers Welfare (formerly Ministry of Agriculture), a branch of the Government of India.
38. NABARD stands for National Bank for Agriculture and Rural Development.
39. NABARD came into existence on 12 July, 1982.
40. Headquarters- Mumbai, Chairman- Dr Harsh Kumar Bhanwala.

Recent Ecological Facts

1. Total forest cover of India 70.827 mh a (2017), which is 24.39 % of Total geographical area
2. The theme of earth day, 2018 was "End Plastic Pollution"
3. Madhya Pradesh has the largest forest cover area
4. On percentage basis Lakshadweep (with 90.33 per cent) has the highest forest area
5. Norway has become the first country to ban deforestation
6. Scientists of UK developed outdoor laboratory named 'Sci-fi' to track how forest trees respond to level of CO_2 concentrations.

Some Important Facts about Irrigation in India and World

1. Punjab (98.1 %)>Haryana (87.6%)>UP (75.9%)
2. Highest area under irrigation state in India: U. P
3. Sugarcane (93.7 %) is highly irrigated crop in India
4. Highest non-irrigated area in India : Mizoram
5. Area under micro irrigation system in India: 7.73Mha (2014-15)
6. Area under drip irrigation in India: 3.37 Mha (Highest area Maharashtra)
7. Area under sprinkler in India (2014-15): 4.35 Mha (Highest in Rajasthan)
8. Sources of irrigation tubewell (57 %), Canals (32 %)
9. USA is having highest micro irrigation system: Sprinkler
10. Over 324 Mha are equipped for irrigation worldwide.
11. 42% of the world irrigation is located in only 2 countries: China and India
12. In 2010, China became the country with the largest irrigation area
13. Total sprinkler irrigation in the world 35 Mha

Transgenic Crops

1. Final approval committee on release of transgenic in India – is GEAC (Genetic Engineering Approval Committee)
2. Leading countries in transgenic use - USA>Brazil > Argentina> Canada> India (5th)
3. India has the world's fifth largest cultivated area under genetically modified (GM) crops,
4. Leading transgenic crops - Soybean >Cotton>Maize > Canola
5. Area under transgenic plants in India (2018)-11.4mha.
6. Area under BT- cotton (below 90% of cotton production): Maharashtra stands first
7. In the past year, the government has approved field trials of GM crops for field trails include transgenic Rice, Cotton, Maize (corn), Mustard, Brinjal and Chickpea.

Fertilizer Consumption Statewise

1. All India Fertilizer consumption in 2016- 130.66 kg/ha.
2. Lowest fertilizer consumption states are Nagaland, Arunachal Pradesh, Sikkim.
3. U.P. is the highest total fertilizer consumption state
4. Imported fertilizer- Potash from Jordan and Tunesia
5. India is third in fertilizer production and second in consumption
6. NPK ratio in 2016-17 was 6.7:2.7:1 (India)
7. Standard NPK for cereals: 4:2:1 and Pulses: 1:2:1
8. Total nutrient consumption (N+P2O5 +K2O) during 2016-17 - 25.95 million metric tonne (MMT).

Pesticide Consumption

1. The agricultural states like Maharashtra and Uttar Pradesh are the toppers in pesticide consumption in India
2. Crop- 1st Paddy (28 %), 2nd Cotton (20 %)

3. Total registered pesticides in India- 290
4. Total banned pesticides in India - 66
5. Central Insecticides Board and Registration Committee (CIB & RC)-To facilitate the registration of safe, efficacious and quality pesticides for domestic use and export.

World Agricultural Scenario and Position of India

- Rice: China > India > Indonesia
- Maize: USA >China >Brazil (India 6th)
- Wheat: Europe Union> China > India >USA
- Groundnut: China > India>Nigeria
- Sugarcane: Brazil > India>China
- Total Cereals: China > USA > India
- Coarse Cereals: USA > China > Brazil > India
- Total Pulses: India – 1st
- Mustard & Rapeseed: China > Canada > India
- Vegetables: China > India> USA
- Fruits & Veg: China > India
- Cotton: China > USA > India
- Tobacco: China > Brazil > India
- Tea, Jute & Allied Fibers: India – 1st
- Coffee-India- 6th
- Cattle Population: 1st India (16.5%)
- Buffalo Population: 1st India (56.7%)
- Milk Production: 1st India (15%)
- Egg Production: China>USA>Japan>India
- Total Agricultural Land in the World: 7.5%
- Total Area of India- 328.7 M Ha-2.4% Of World-7th Position
- Total Arable Land in India- 162 M Ha-2nd After USA
- Lowest Agricultural Based Country: U.K. (6%)
- Highest Agricultural Based Country: Malavi (92%)
- Total Irrigated Area- 58 M Ha-21% Of World-1st Position

Agricultural Commodity Top Producer State*

1. Largest Total Food Grains producing state in India is – Uttar Pradesh
2. Largest Rice producing state in India is – West Bengal
3. Largest Wheat producing state in India is – Uttar Pradesh
4. Largest Maize producing state in India is - Andhra Pradesh
5. Largest Coarse Cereal producing state is - Karnataka, Rajasthan
6. Largest Oilseeds producing state in India is - Gujarat
7. Largest Rapeseed & Mustard producing state in India is - Rajasthan
8. Largest Soyabean producing state – Madhya Pradesh
9. Largest Sunflower producing state in India is - Karnataka
10. Largest Sugarcane producing state in India is - Uttar Pradesh
11. Largest Sugar producing state in India is – Maharashtra
12. Largest Cotton producing state in India is - Gujarat
13. Largest Jute producing state in India is - West Bengal
14. Largest Horticulture Products producing state in India is - West Bengal
15. Largest Vegetables producing state in India is - West Bengal
16. Largest Fruits producing state in India is - Andhra Pradesh
17. Largest Loose Flowers producing state in India is - Tamil Nadu
18. Largest Cut Flowers producing a state in India is - West Bengal
19. Largest Total Spices producing state in India is – Andhra Pradesh
20. Largest Total Plantation Crops producing state in India is - Tamil Nadu
21. Largest Banana producing state in India is - Tamil Nadu
22. Largest Guava producing state in India is - Madhya Pradesh
23. Largest Grapes producing state in India is - Maharashtra
24. Largest Apple producing state in India is - Jammu & Kashmir
25. Largest Areca Nut producing state in India is - Karnataka
26. Largest Cocoa producing state in India is - Kerala
27. Largest Coconut producing state in India is - Tamil Nadu

28. Largest Cashew nut producing state in India is - Maharashtra
29. Largest Citrus, Lemon, Mosambi producing state in India is - Andhra Pradesh
30. Largest Orange producing state in India is - Punjab
31. Largest Litchi producing state in India is - Bihar
32. Largest Mango producing state in India is - Uttar Pradesh & Andhra Pradesh
33. Largest Pineapple producing state in India is - West Bengal
34. Largest Brinjal producing state in India is – Odisha

First Advance Estimates of Production of Commercial Crops for 2018-19

S. No.	Crop	Targets (Lakh Tonnes)	1st Adv. Est. (Lakh Tonnes)
1.	**Rice**	113.0	99.24
2.	**Wheat**	100.00	-
3.	**Jowar**	2.10	1.88
4.	**Bajra**	9.50	7.77
5.	**Ragi**	2.30	1.68
6.	**Small millet**	0.60	0.33
7.	**Maize**	27.30	21.47
8.	**Coarse cereal**	46.70	33.13
9.	**Cereals**	259.70	132.37
10.	**Tur**	4.50	4.08
11.	**Gram**	10.5	-
12.	**Urad**	2.80	2.65
13.	**Moong**	2.10	1.58
14.	**Total pulse**	24.00	9.22
15.	**Total food grain**	283.70	141.59
16.	**Groundnut**	75.40	63.28
17.	**Rapeseed**	84.86	-
18.	**Soyabean**	148.20	134.59
19.	**Total Nine Oilseeds**	359.99	221.89
20.	**Sunflower**	3.02	0.94
21.	**Sugarcane**	3550	3838.92
22.	**Cotton**	355.00	324.83
23.	**Jute**	105.00	96.65
24.	**Sesamum**	10.07	7.10

Animal Products Statistics

Commodity	Total Production (per year)	Per Capita Availability	ICMR Recommendations
Milk	165.4 MT	355 grams/day	280 grams/day
Meat	7.4 MT	2.96 kg/year	11 kg/year
Eggs	88.1 billion	69 eggs/year	182 eggs/year

Positions of India in the World

Ranking	Sector
1st	Total Livestock Population, Milk Production, Cattle Population, Buffalo Population, Carabeef Production, Goat Milk Production, Total Bovine Population
2nd	Goat Population, Bristle Production (a pig industry by-product), Fish Production
3rd	Sheep Production, Egg Production
4th	Chicken Production
5th	Poultry Meat Production, Poultry Production
8th	Duck Production
9th	Camel Population, Wool Production

Chief States in India for Livestock Production

Feature	State
Highest milk producing state	Uttar Pradesh (25,198 L)
State with highest per capita availability of milk	Punjab (1075 g)
State with highest per capita availability of eggs	Andhra Pradesh (312 eggs)
Highest meat production	Uttar Pradesh
Highest goat meat production	West Bengal
Highest poultry and sheep meat production	Andhra Pradesh
Highest wool production	Rajasthan
Highest indigenous milch cows	Madhya Pradesh
Highest Cross-bred or exotic cows	Tamil Nadu
Highest number of poultry birds	Andhra Pradesh
Highest number of mules	Jammu & Kashmir
Highest number of live stock	Uttar Pradesh
Highest cattle population	Madhya Pradesh

Highest buffalo population	Uttar Pradesh
Highest sheep population	Andhra Pradesh
Highest goat population	Rajasthan
Highest pig population	Assam
Highest camel population	Rajasthan
Highest mithun population	Arunachal Pradesh
Highest yak population	Jammu & Kashmir
Highest horses and ponies population	Uttar Pradesh
Highest donkey population	Rajasthan

Livestock Census

- First livestock census in India was conducted as Dairy Cattle Census in 1919.
- Livestock Census, 2012 was 19th Livestock Census conducted with reference date of 15th October, 2012.
- Livestock Census takes place at every five years (Quinquennial)
- Total livestock population: 512.05 million
- Total poultry population: 729.02 million
- Milch animals: 118.59 million
- Total bovine population: 299.99 million
- Total cattle population: 190.90 million
- Total indigenous cattle: 151.17 million
- Total buffalo population: 108.70 million
- Total sheep population: 65.06 million
- Total goat population: 135.17 million
- Total pig population: 10.29 million
- Total camel population: 0.4 million
- Total horses and ponies population: 0.62 million
- Total mules population: 0.19 million
- Total donkey population: 0.32 million
- Total yak population: 0.07 million

- Total mithun population: 0.29 million
- Total elephant population: 22,000
- Total rabbit population: 0.59 million
- Number of animals under Livestock Insurance Scheme of DADF: 1017713 (2015-16)
- Share of Uttar Pradesh in Total Milk Production in India: 16.8%
- Share of Tamil Nadu in Total Egg Production in India: 18.9%
- Share of Uttar Pradesh in Total Meat Production in India: 18.23%
- Share of Rajasthan in Total Wool Production in India: 32.9%
- 18.5 per cent of Total World Milk is produced in India

Specieswise Milk Contribution to Total Milk Production in India

Species	**Percentage of Total Milk**
Buffalo indigenous	35 per cent
Buffalo non-descript	14 per cent
Cow indigenous	11 per cent
Cow non-descript	9 per cent
Cow cross-bred	26 per cent
Cow exotic	1 per cent
Goat	3 per cent

Specieswise Egg Contribution to Total Egg Production in India

Species	**Percentage of Total Eggs**
Improved fowl	86.80 per cent
Desi fowl	11.96 per cent
Desi duck	0.96 per cent
Improved duck	0.28 per cent

Specieswise Meat Contribution to Total Meat Production in India

Species	**Percentage of Total Meat**
Poultry	47.05 per cent
Buffalo	19.80 per cent
Goat	14.25 per cent
Sheep	7.68 per cent
Pig	6.50 per cent
Cattle	4.72 per cent

Categorywise Contribution to Total Wool Production in India

Category	Percentage of Total Meat
Ewe	68.35 per cent
Ram	16.71 per cent
Lamb	14.94 per cent

Percentage of the World Livestock Present in India

Species	Percentage present in India
Total Livestock	11.54 per cent
Buffalo	56.7 per cent
Small ruminants	20.4 per cent
Cattle	12.5 per cent
Poultry	3.1 per cent

World Leaders in Livestock Production

Category	Number One in the World
Bovines	India
Cattle	India
Buffalo	India
Milk production	India
Carabeef	India
Pigs	China
Donkeys	China
Sheep	China
Goat	China
Duck	China
Fish	China
Eggs	China
Chevon	China
Mutton	China
Pork	China
Camels	Somalia
Poultry	USA
Poultry meat	Brazil
Fibre	Australia
Wool	Australia
Beef	USA

Percentage of the World Livestock Present in India

Species	Percentage present in India
Total Livestock	11.54 per cent
Buffalo	56.7 per cent
Small Ruminants	20.4 per cent
Cattle	12.5 per cent
Poultry	3.1 per cent

Share of Agriculture & Allied and Livestock Sector in GVA – (At current prices in Rs. Crore)

Year	GVA (Total)	GVA (Agriculture & Allied)		GVA (Livestock Sector)	
2015-16	12458642	2175547	17.5 %	560613	4.5 %

Fish Production Statistics

- Total Fish Production in India: 10.07 MT
- Fish Seed Production: 43266 million fry

Inland Fisheries Resources

- Total Water Bodies: 73.12 Lakh Ha

Marine Fisheries Resources

- Approximate length of coastline: 8118 km
- Continental Shelf: 5.30 Lakh sq.km.
- Number of Landing Centers: 1537
- Number of Fishing Villages: 3432
- Number of fisherman families: 874749
- Total fishing crafts: 221450.

Chapter 3

Schemes

Govt. Schemes

National Seed Research and Training Centre (NSRTC), Varanasi:

The National Seed Research and Training Centre (NSRTC), Varanasi (Uttar Pradesh) has been functioning since October, 2005 and has been notified as Central Seed Testing and Referral Laboratory (CSTL) with effect from 01.04.2007.

Seed Village Programme

In order to upgrade the quality of farmer saved seeds which is about 60-65% of the total seeds used for crop production programme, following interventions are made:

a) 50% assistance for the seeds for *cereal crops* and *60% for oilseeds, pulses, fodder and green manure crops* is provided for distribution of foundation/certified seeds required for one acre area per farmer.

b) Farmers' Trainings : Financial assistance of Rs.15000 per group (50-150 farmers each group) is provided for farmers training on seed production and post-harvest seed technology (Rs. 0.15 lakh)

c) Seed treating/dressing drums : Financial assistance for treating seeds produced in the Seed Village is available @ 3500 per seed treating drum of 20Kg capacity and Rs. 5000 per drum of 40Kg capacity.

d) Seed Storage bins : To encourage farmers to develop storage capacity of appropriate quality, financial assistance is available to farmers for purchasing seed Storage bins. The rate of assistance is as under: Average farm power availability for the cultivated areas of the country has been increased from 0.48 kW/ha in 1975-76 to 1.84 kW/ha in 2012-13 and expected to cross 2.02 kW/ha by 2017-18

Pradhan Mantri Fasal Bima Yojna (PMFBY)

Under the PMFBY, a uniform maximum premium of only **2%** is paid by farmers for all Kharif crops and **1.5%** for all Rabi crops. In case of annual commercial and horticultural crops, the maximum premium to be paid by farmers is only **5%**.

Integrated Scheme for Agricultural Marketing (ISAM)

a) Marketing Division is also implementing ongoing Central Sector Schemes from **XIIth Plan**, which have been integrated into a new scheme viz. the Integrated Schemes for Agricultural Marketing (ISAM) w.e.f. 01.04.2014. During 2017-18, National Agriculture Market Scheme popularly known as e-NAM scheme has also been made part of it. Agricultural Marketing Infrastructure (AMI) sub-scheme of Integrated Scheme for Agricultural Marketing (ISAM) is being implemented since 1st April, 2014. The erstwhile two schemes of DAC & FW

i) Gramin Bhandaran Yojna (GBY) implemented since 1st April, 2001 and

ii Scheme for Development/Strengthening of Agricultural Marketing Infrastructure, Grading and Standardization (AMIGS) implemented since 20th October, 2004 are subsumed into one sub-scheme known as Agricultural Marketing Infrastructure (AMI).

Marketing Research and Information Network (MRIN)

a) This scheme was launched as Central Sector Scheme in March, 2000 with objective to collect and disseminate information on price, arrival and other market related data for the benefit of farmers and other market users.

Small Farmers' Agribusiness Consortium (SFAC)

SFAC was set up as a registered society on 18 January, 1994.

Sub-Schemes Launched under RKVY

Bringing Green Revolution to Eastern India (BGREI)

Initiated in 2010-11, BGREI has targeted productivity improvement in the rice-based cropping system of Assam, West Bengal, Odisha, Bihar, Jharkhand, Eastern Uttar Pradesh and Chhattisgarh.

Saffron Mission

The Scheme was initiated in 2010-11 for revival of saffron cultivation in Jammu & Kashmir.

Crop Diversification Programme (CDP)

Pursuant to an announcement in the Union Budget for 2013-14, this scheme had been launched to promote technological innovation to encourage farmers to choose crop alternatives.

Reclamation of Problem Soil (RPS)

This programme was initiated as a sub-scheme of RKVY during 2016-17 for enhancing soil fertility & productivity for meeting the demand for food grains in the country.

Foot and Mouth Disease Control Programme (FMD-CP)

This programme was initiated as a sub-scheme of RKVY during 2016-17 for supplementing the efforts of the State Governments for prevention, control and containment of this animal disease.

Additional Fodder Development Programme (AFDP)

This programme was launched in 2014-15 in order to mitigate the impact of drought on livestock. This programme was initially launched as the Accelerated Fodder Development Programme during 2011- 12 to accelerate the production of fodder through intensive promotion of technologies to ensure availability of fodder throughout the year.

Scheme for Drought Affected Area

This scheme has been initiated in 2017- 18 with an allocation of Rs. 25 crores for a pilot implementation of interventions to address the issue of the most vulnerable drought prone districts in the country.

Peri-Urban horticulture

This scheme has been initiated in 2017-18 with the objectives of greening cities, utilizing waste, meeting the needs of food and nutrition and serving the environment by promoting cultivation of fruits, flowers, vegetables medicinal and aromatic plants in selected cities.

Swachh Bharat : This scheme has been initiated in 2017-18

RKVY-RAFTAAR

The scheme has been approved by Union Cabinet on 1st November, 2017 for continuation upto March, 2020 as Rashtriya Krishi Vikas Yojna - Remunerative Approaches for Agriculture and Allied sector Rejuvenation (RKVY-RAFTAAR) with an outlay of Rs. 15722.00 crore for 2017-18 to 2019-20.

National Food Security Mission (NFSM)

NFSM was launched in 2007-08 to increase the production of rice, wheat and pulses by 10, 8 and 2 million tonnes, respectively by the end of 11th Plan through area expansion and productivity enhancement; restoring soil fertility and productivity; creating employment opportunities; and enhancing farm level economy. The Mission was continued during 12th Plan with new target of additional production of 25 million tonnes by the end of 12th Plan including coarse cereal crops.

National Horticulture Mission (NHM)

This Centrally Sponsored Scheme was launched in the year 2005-06 aims at the holistic development of horticulture sector.

National horticulture Board

The National Horticulture Board (NHB) was established in the year 1984 by the Government of India as an autonomous organization and registered as a society with the Registrar of Firms and Societies, Chandigarh with its headquarters in Gurugram. Interventions in horticulture in the country, have led to increase per capita availability of fruits from 158 gm/person/day in 2007-08 to 201 gm/person/day in 2016-17. Similarly, per capita availability of vegetables has increased from 309 gm/person/day in 2007-08 to 385 gm/person/day in 2016-17.

National Agro-forestry & Bamboo Mission (NABM)

National Bamboo Mission (NBM) is a Centrally Sponsored Scheme of DAC&FW w.e.f. 2006-07 across the country except Haryana. It was subsumed under Mission for Integrated Development of Horticulture (MIDH) during 2014-15 and renamed as National Agro-forestry & Bamboo Mission (NABM) in July, 2015.

Coconut Development Board

Coconut Development Board Kochi, (CDB) is a statutory body established by Govt. of India by an Act of Parliament (Coconut Development Board Act, 1979) and came in to existence in January, 1981. India being the largest producing country in the world occupies a 31% of global production.

S. No	Scheme	Ministry	Date of Launch	Objective
1.	Pradhan Mantri Jan-Dhan Yojna (PMJDY)	Ministry of Finance	28 Aug, 2014	It is a National Mission for Financial Inclusion of all households to ensure access to financial services, namely, Banking/ Savings & Deposit Accounts, Remittance, Credit, Insurance, Pension in an affordable manner.
2.	Kisan Vikas Patra (KVP) Re-introduced	Government of India	2014	**Kisan Vikas Patra** is a saving certificate scheme which was first launched in 1988 by India Post . Kisan Vikas Patra is an instrument on which fixed interest is paid by the Govt. This instrument is sold by the Govt. and can be purchased through any Post Office. Kisan Vikas Patra can be purchased by making payment either in cash/ cheque/ pay order/ demand draft.
3.	Pradhan Mantri Sukanya Samriddhi Yojana (PMSSY)	Ministry of Women and Child Development	22 Jan, 2015	The scheme launched as a part of the *"Beti Bachao, Beti Padhao"* campaign targeted at the parents of girl children. The scheme encourages parents to build a fund for the future education and marriage expenses for their female child.
4.	Pradhan Mantri MUDRA Yojna (PMMY) (MUDRA-Micro Units Development and Refinance Agency)	Ministry of Finance	8 April, 2015	The purpose of MUDRA is to provide funding to the non-corporate small business sector. The small businesses/startups or entrepreneurs can avail loans from Rs. 50 thousand to 10 Lakh to start/grow their business under the three, Shishu, Kishore and Tarun categories of the scheme.

5.	**Pradhan Mantri Jeevan Jyoti Bima Yojna**	**Ministry of Finance**	**9 May, 2015**	**Pradhan Mantri Jeevan Jyoti Bima Yojna is a government-backed Life insurance scheme in India. Its main objective is to provide life insurance cover to all Indian citizens.**
6.	Pradhan Mantri Suraksha Bima Yojna (PMSBY)	Ministry of Finance	9 May, 2015	Pradhan Mantri Suraksha Bima Yojna is a government-backed accident insurance scheme in India. Its main objective is to provide accidental insurance cover to all Indian citizens.
7.	Atal Pension Yojna (APY)	Ministry of Finance	9 May, 2015	Atal Pension Yojna (previously known as Swavalamban Yojna) is a government-backed pension scheme in India targeted at the unorganized sector. The scheme is targeted to the private unorganized sector and is open to all Indian citizens between the age of 18 to 40 years.
8.	Pradhan Mantri Awas Yojna (PMAY)	Ministry of Housing and Urban Affairs	25 June, 2015	The scheme envisages the vision of housing for all by the year 2022. Under this scheme, interest subsidy on purchase, construction, extension, improvement of house is provided to customers belonging to Economical Weaker Section, Lower Income Group, Middle Income Group.
9.	NITI Aayog (The National Institution for Transforming India)	Govt of India	1 Jan, 2015	It replaced the Planning Commission. The aim of NITI Aayog's creation is to foster involvement and participation in the economic policy-making process by the State Governments of India.
10.	**Rashtriya Gram Swaraj Abhiyan**	**Ministry of Panchayati Raj**	**NA**	**It is an updated version of Rajiv Gandhi Panchayat Sashaktikaran Abhiyan. It will strengthen the Panchayati Raj system across the country and address critical gaps that constrain its success.**
11.	National Food Security Mission (NFSM)	GOI	October, 2007.	The National Food Security Mission (NFSM) is being implemented with the new target of additional production of 25 million tonnes of food grains comprising 10 million tonnes rice, 8 million tonnes wheat, 4 million tones pulses, and 3 million tonnes coarse

				the end of the Twelfth Five Year Plan (2016-17). The revamped NFSM is being implemented from 2014-15 in 619 districts of 28 states. To promote the use of bio-fertilizers, subsidy on bio-fertilizer has also been enhanced from Rs. 100 per ha to Rs.300 per ha.
12.	National Horticulture Mission (NHM)	GOI	July 8, 2004	The NHM's key objective is to develop horticulture to the maximum potential available in the state and to augment production of all horticultural products (fruits, vegetables, flowers, coco, cashew nut, plantation crops, spices, medicinal aromatic plants) in the state.
13.	Restructured National Bamboo Mission (NBM)	Ministry of Agriculture	25 April, 2018	The Mission envisages promoting holistic growth of bamboo sector by adopting area-based, regionally differentiated strategy and to increase the area under bamboo cultivation and marketing. Under the Mission, steps have been taken to increase the availability of quality planting material by supporting the setting up of new nurseries and strengthening of existing ones. To address forward integration, the Mission is taking steps to strengthen marketing of bamboo products, especially those of handicraft items.
14.	The mission for Integrated Development of Horticulture (MIDH)	Ministry of Agriculture	2014-15	The mission for Integrated Development of Horticulture (MIDH) is a Centrally Sponsored Scheme for the holistic growth of the horticulture sector covering fruits, vegetables, root & tuber crops, mushrooms, spices, flowers, aromatic plants, coconut, cashew, cocoa and bamboo. Under MIDH, Government of India (GOI) contributes 60%, of total outlay for developmental programmes in all the states except states in North-East and Himalayas, 40% share is contributed by State Governments. In the case of North-Eastern States and Himalayan States, GOI contributes 90%.

15.	**Krishi Dak**	**IARI**	**2015**	**IARI initiated this novel scheme in 20 districts in which postmen supplied seeds of improved varieties of crops to the farmers in far-flung areas. Owing to its success and popularity, this scheme is being extended in 100 districts of 14 states with the association of Krishi Vigyan Kendras. This will provide improved seed to farmers at their doorstep.**
16.	Rashtriya Gokul Mission	Department of Animal Husbandry and Dairying and Fisheries	December, 2014	Rashtriya Gokul Mission (RGM) has been launched in December, 2014 with an outlay of Rs. 2025 crore for development and conservation of indigenous breeds through selective breeding in the breeding tract and genetic upgradation of non-descript bovine population. The scheme comprises of two components, namely National Programme for Bovine Breeding (NPBB) and National Mission on Bovine Productivity (NMBP).
17.	Mera Gaon, Mera Gaurav			This scheme is being launched involving agricultural experts of agricultural universities and ICAR institutes for effective and deeper reach of scientific farming to the villages. A group of experts will be associated with one particular village to create awareness and adoption of new technologies including farm investment, loans, availability of inputs and marketing. All the scientists from ICAR and agricultural universities will participate in this initiative.
18.	Kisan Vikas Patra	Ministry of Finance	3 March, 2015 (Re-launched)	Kisan Vikas Patra is a saving certificate scheme which was first launched in 1988 by India Post based on the recommendations of Shyamla Gopinath Committee Govt of India closed the scheme because of its flaws. Its aim is to provide safe and secure investment avenues to the small investors.
19.	Soil Health Card Scheme	Ministry of Agriculture **Farmers Welfare**	17 Feb, 2015	Under this Scheme, Government issues soil cards to farmers which will carry crop-wise recommendations of nutrients and **fertilizers required for the individual farms to help farmers to improve productivity through judicious use of inputs.**
20.	Digital India	Ministry of Electronics and Information Technology	2 July, 2015	To ensure that Govt services are made available to citizens electronically by improved online infrastructure and by increasing Internet connectivity or by making the country digitally empowered in the field of technology.
21.	Pradhan Mantri Kaushal Vikas Yojna (PMKVY)	Ministry of Skill Develop-ment & Entrepre-neurship	16 July 2015	To encourage aptitude towards employable skills and to increase working efficiency of probable and existing daily wage earners, by giving monetary awards and rewards and by providing quality training to them.

22.	Mission Indradhanush	Ministry of Health & Family Welfare	25 Dec, 2014	It aims to immunize all children under the age of 2 years, as well as all pregnant women, against seven vaccine preventable diseases. The diseases being targeted are diphtheria, whooping cough, tetanus, poliomyelitis, tuberculosis, measles and hepatitis B.
23.	Deen Dayal Upadhyaya Gram Jyoti Yojna (DDUGJY)	Ministry of Power	25 July, 2015	The scheme designed to provide continuous power supply to rural households and adequate power to agricultural consumers. It replaced Rajiv Gandhi Gramin Vidyutikaran Yojna (RGGVY).
24.	Deen Dayal Upadhyaya Gramin Kaushalya Yojna (DDUGKY)	Ministry of Rural Develop-ment	25 Sept, 2014	The Vision of DDU-GKY is to "Transform rural poor youth into an economically independent and globally relevant workforce by developing skills and providing them with jobs having regular monthly wages.
25.	Sansad Adarsh Gram Yojna (SAGY)	Ministry of Rural Develop-ment	11 Oct, 2014	It focuses on development of model villages called "Adarsh Gram". The programme was launched by the Prime Minister of India, Narendra Modi on the birth anniversary of Jayaprakash Narayan, on 11 October, 2014.
26.	Pradhan Mantri Fasal Bima Yojna (PMFBY)	Ministry of Agriculture & Farmers Welfare	11 Oct, 2014	Its main objective is to provide insurance cover to rabi and kharif crops and financial support to farmers in case of damage of crops. It replaced NAIS and MNAIS schemes.
27.	Pradhan Mantri Gram Sinchai Yojna (PMGSY)	Ministry of Agriculture & Farmers Welfare	01 July, 2015	The motive of Pradhan Mantri Gram Sinchai Yojna will be to take irrigation water to each and every agricultural field in the country and boost agricultural yield
29.	Pradhan Mantri Jan Aushadhi Yojna (PMJAY)	Dept. of Pharmace uticals, Govt. Of India	1 July 2015	Its aim is to provide quality medicines at affordable prices to the masses through special kendra?s known as Pradhan Mantri Bhartiya Jan Aushadhi Kendra. PMBJK has been set up to provide generic drugs, which are available at lesser prices but are equivalent in quality and efficacy as expensive branded drugs.
30.	Make in India	Ministry of Finance	25 Sept, 2014	To encourage national, as well as multi-national companies to manufacture their products in India and create jobs and skill enhancement in 25 sectors.
31.	Swachh Bharat Abhiyan	Ministry of Drinking Water & Sanitation in Rural areas; Ministry of Housing & Urban Affairs in Urban areas	2 Oct, 2014	To keep the streets, roads and infrastructure of the country's 4,041 statutory cities and towns and its rural areas clean and fulfill Mahatma Gandhi"s dream of a clean and hygienic India.

32.	**Pandit Deendayal Upadhyay Shramev Jayate Yojana (PDUSJY)**	**Ministry of Labour and Employment**	**16 Oct, 2014**	**Objective of this scheme is to create conducive environment for industrial development and doing business with ease and also expanding government support to impart skill training for workers.**
33.	Atal Mission for Rejuvenation and Urban Transformation (AMRUT)	Ministry of Urban Development	24 June, 2015	Providing basic services (water supply, sewerage, urban transport) to households & build Amenities in cities which will improve the quality of life for all, especially the poor and the disadvantaged.
34.	Swadesh Darshan Yojna (SDY) & Pilgrimage Rejuvenation and Spiritual Augmentation Drive (PRASAD)	Ministry of Tourism	09 March, 2015	To provide Central Financial Assistance (CFA) to State Governments/Union Territory Administrations, including the places of religious importance, for the development of various tourism projects and tourism infrastructure in the country. Under PRASAD 12 cities have been identified for development.
35.	Rashtriya Gokul Mission	Ministry of Agriculture & Farmers Welfare	16 Dec, 2014	The "Rashtriya Gokul Mission" aims to conserve and develop indigenous bovine breeds in a focused and scientific manner
36	Pradhan Mantri Sansad Adarsh Gram Yojna	Minister of Rural Development	11 Oct, 2014	PMSAGY is a rural development programme broadly focusing upon the development in the villages which includes social development, cultural development and spread motivation among the people on the social mobilization of the village community. The programme was launched by the Prime Minister of India, Narendra Modi on the birth anniversary of Jayaprakash Narayan, on 11 October, 2014.

Long Form of Govt. Schemes and Yojnas

1. **UDAY**: Ujwal Discom Assurance Yojna
2. **PMMY**: Pradhan Mantri Mudra Yojna
3. **PMJDY**: Pradhan Mantri Jan Dhan Yojna
4. **PMJJBY:** Pradhan Mantri Jeevan Jyoti Bima Yojna
5. **PMSBY**: Pradhan Mantri Suraksha Bima Yojna
6. **APY**: Atal Pension Yojna
7. **KVP**: Kisan Vikas Patra
8. **SBA:** Swachh Bharat Abhiyan
9. **PMSAGY**: Pradhan Mantri Sansad Adarsh Gram Yojna

10. **AMRUT**: Atal Mission for Rejuvenation & Urban Transformation
11. **NGM**: Namami Ganga Yojna
12. **HRIDAY**: Heritage City Development & Augmentation Yojna
13. **MUDRA**: Micro Units Development & Refinance Agency
14. **SETU**: Self Employment & Talent Utilization
15. **NPS**: National Pension Scheme
16. **PMKVY**: Pradhan Mantri Kaushal Vikas Yojna
17. **PMKSY**: Pradhan Mantri Krishi Sinchai Yojna
18. **BBBP YOJANA**: Beti Bachao, Beti Padhao Yojna
19. **SSY:** Sukanya Samriddhi Yojna
20. **PMFBY**: Pradhan Mantri Fasal Bima Yojna
21. **PMGSY**: Pradhan Mantri Gram Sadak Yojna
22. **PMUY**: Pradhan Mantri Ujjwala Yojna
23. **PMGKY**: Pradhan Mantri Garib Kalyan Yojna
24. **DICGC**: Deposit Insurance and Credit Guarantee Corporation
25. **TEC INDIA:** Transform Energise and Clean India
26. **PACS:** Primary Agriculture Credit Societies
27. **CPI:** Consumer Price Index
28. **WPI**: Wholesale Price Index
29. **CAD**: Current Account Deficit
30. **KVKs:** Krishi Vigyan Kendras
31. **MSMEs:** Micro, Small and Medium Enterprises
32. **CBS**: Core Banking Solution
33. **CORE**: Centralized Online Real Time Exchange
34. **LTIG**: Long Term Irrigation Fund
35. **MIF:** Micro Irrigation Fund
36. **NAM:** National Agricultural Market
37. **DIDF**: Dairy Processing and Infrastructure Development Fund

38. **MGNREGA**: Mahatma Gandhi National Rural Employment Guarantee Act
39. **PMAY**: Pradhan Mantri Awaas Yojna
40. **NRDWP**: National Rural Drinking Water Programme
41. **SWAYAM**: Study Webs of Active Learning for Young Aspiring Minds
42. **PMKK**: Pradhan Mantri Kaushal Kendra
43. **SANKALP:** Skill Acquisition and Knowledge Awareness for Livelihood Promotion Programme
44. **STRIVE**: Skill Strengthening for Industrial Value Enhancement
45. **MSK**: Mahila Shakti Kendra
46. **NHB:** National Housing Bank
47. **RRSK**: Rashtriya Rail Sanraksha Kosh
48. **M-SIPS:** Modified Special Incentive Package Scheme
49. **EDF**: Electronic Development Fund
50. **TIES**: Trade Infrastructure for Export Scheme.

❑❑❑

Chapter 4

Current Agriculture News 2018

GDP Growth Soars To 8.2% In 1st Quarter of Current Fiscal

i) Indian economy grew by **8.2%** in the first quarter (April-June) of the current fiscal, the highest in over two years. This announcement was made by Finance Minister, Arun Jaitley. The manufacturing sector grew by **13.5%,** which signals very good turnaround in the sector.

ii) In a related development Reserve Bank of India stated Non-food credit growth in the system accelerated to 10.6% for July as compared to previous year, driven by loans to the services sector growing at the faster clip.

iii) Credit to the services sector grew **23%** for the reporting period, up from the year-ago period's **4.9%.**

Suresh Prabhu Launches Technology Initiatives for Coffee Stakeholders

i) Minister for Commerce & Industry and Civil Aviation, Suresh Prabhu launched ***Coffee Connect*** - India coffee field force app and ***Coffee Krishi Tharanga*** - digital mobile extension services for coffee stakeholders at a function in New Delhi.

ii) The mobile app Coffee Connect has been developed to ease the work of field functionaries and to improve the work efficiency. The Coffee Krishi Tharanga services are aimed at providing customized information and services to increase productivity, profitability, and environmental sustainability.

Sanctioned Rs. 335 Crore to West Bengal in August: NABARD

i) National Bank for Agriculture and Rural Development (NABARD) has sanctioned an amount of **Rs 335** crore under the Rural Infrastructure Development Fund (RIDF) to West Bengal. The fund, allotted in August, would be used for 158 minor irrigation projects and 23 flood protection projects.

ii) The minor irrigation projects are expected to benefit **20,506 hectares**, covering **3.09 lakh** population in **699 villages** of **22 districts**, while the flood protection projects would address the problem of rapid bank erosion caused by rivers.

Haryana Signs MoU with Indian Oil for Ethanol Plant

i) The Haryana government signed an MoU with **Indian Oil Corporation (IOC)** to set up an ethanol plant in Panipat with an outlay of over **Rs. 900 crore.**

ii) The initiative with a proposed capacity of **100 kilolitres** of ethanol per day will manage crop residue and prevent straw burning before the upcoming paddy season. The pact will be valid for one year.

Andhra Launches e-Rythu To Boost Cashless Farm Ecosystem

i) The government of Andhra Pradesh launched a mobile platform, e-Rythu (e-farmer in Telugu) which will enable small scale farmers to market their produce at reasonable prices.

ii) e-Rythu was launched by Chief Minister N. Chandrababu Naidu. The platform has been developed by Mastercard Labs for Financial Inclusion in Nairobi, Kenya, and has been customized for India use by the Labs team based at Pune.

Cabinet Approves New Umbrella Scheme 'Pradhan Mantri Annadata Aay Sanrakshan Abhiyan'

i) The Union Cabinet approved new Umbrella Scheme 'Pradhan Mantri Annadata Aay Sanrakshan Abhiyan, PM-AASHA.

ii) The Scheme is aimed at ensuring remunerative prices to the farmers for their produce as announced in the Union Budget for 2018, the announcement was made by the Agriculture Minister Radha Mohan Singh in New Delhi. The Cabinet has decided to give additional government guarantee of **Rs.16,550 crore** making it Rs. **45,550 crores** in total.

iii) The new Umbrella Scheme includes the mechanism of ensuring remunerative prices to the farmers and is comprised of:

1. Price Support Scheme (PSS),
2. Price Deficiency Payment Scheme (PDPS), and
3. Pilot of Private Procurement and Stockist Scheme (PPPS).

Agri-economist & Padma Bhushan Laureate Vijay Shankar Vyas Passes Away

i) Renowned agricultural economist and former PM's Economic Advisory Committee Member, Professor Vijay Shankar Vyas died after a brief illness. He was 87. He was the Director of the Indian Institute of Management (IIM), Ahmedabad, and the Institute of Development Studies (IDS), Jaipur.

ii) Vyas was a Senior Advisor in the World Bank's agriculture and rural development department. He was also a member of the PM's Economic Advisor Committee.

NABARD Sanctions Rs. 772 crore for Water Projects in Haryana

i) According to the State Minister for Public Health Engineering Banwari Lal, National Bank for Agriculture and Rural Development (NABARD) has sanctioned Rs. **772 crore** for eight water projects in the rural areas of Haryana.

ii) Works on these projects, which will supply water to **256 villages** and **72 hamlets**, are in progress. The state government has sent three more project proposals, costing **Rs. 236.85 crore**, to NABARD for financial assistance.

Nikhil Nanda Appointed as Chairman and MD of Escorts

i) Farm and construction equipment maker Escorts appointed Nikhil Nanda as the Chairman and Managing Director (MD) of the company.

ii) He succeeded his father Rajan Nanda who passed away earlier this week after a brief illness.

Government, NABARD and NWDA Sign Revised MoA for Funding of Projects Under PMKSY

i) The Union Ministry of Water Resources, the NABARD and the National Water Development Agency (NWDA) signed a revised Memorandum of Agreement (MoA) for funding of central share of 99 prioritised irrigation projects under Pradhan Mantri Krishi Sinchayi Yojna (PMKSY).

ii) The funding will be done through the Long-Term Irrigation Fund (LTIF). The revised MoA will enable the Ministry of Water Resources to release Central Assistance to prioritised projects under PMKSY as per the requirements from time to time.

MoU Signed Between TRIFED and National Medicinal Plant Board

i) Tribal Co-operative Marketing Development Federation of India Ltd. (TRIFED) under the Ministry of Tribal Affairs & National Medicinal Plants Board (NMPB), Ministry of AYUSH have signed an MoU together.

ii) The MoU aims to promote Medicinal and Aromatic Plants (MAPs) forest produced for livelihood development of tribal people by promoting primary level value addition to MAPs produce from forest area at the grassroots level.

Khangchendzonga Becomes 11th Biosphere Reserve To be Included in WNBR

i) The Khangchendzonga Biosphere Reserve (in Sikkim) has become the 11th Biosphere Reserve from India that has been included in the UNESCO designated World Network of Biosphere Reserves (WNBR).

ii) The decision to include Khangchendzonga Biosphere Reserve in WNBR was taken at the 30th Session of International Co-ordinating Council (ICC) of Man and Biosphere (MAB) Programme of UNESCO held at Palembang, Indonesia.

Important Cabinet Approvals: 9th August, 2018

i) The World Bio-fuel Day is observed every year on August 10 to create awareness about the importance of non-fossil fuels as an alternative to conventional fossil fuels. The World Bio-fuel Day is being observed by the Ministry of Petroleum & Natural Gas.

ii) On the occasion, PM of India, Narendra Modi unveiled a Booklet on ***"National Policy on Biofuels, 2018"***, and launched ***"Pro-Active and Responsive facilitation by Interactive and Virtuous Environmental Single Window Hub" [PARIVESH]***.

Government Provides Fund of Rs 66000 crore To Increase Country's Forest Cover

i) The Union Government has unlocked green funds worth over **Rs. 66,000 crore** to increase the country's forest cover.

ii) The fund is an accumulated amount of money, which user agencies have been depositing as compensation for diverting forest land for non-forest purposes, including industries and infrastructure, since the past 10 years.

Ministry of Agriculture Proposes to UNFAO To Declare 2019 as 'International Year of Millets'

i) The Union Minister of Agriculture and Farmers' Welfare, Radha Mohan Singh urged to Jose Graziano da Silva, the Director General of the United Nations Food and Agriculture Organisation (UNFAO) to declare the upcoming year as the 'International Year of Millets'.

ii) The adoption of this proposal by the FAO with the support of its Member Nations will enable it to be moved to the United Nations General Assembly (UNGA) for the declaration of an upcoming year as International Year of Millets.

FSSAI Constitutes B Sesikeran Committee to Review Food Labelling Standards

i) The Food Safety and Standards Authority of India (FSSAI) constituted a group of experts from the health and nutrition sector to look into the issue of food labelling.

ii) The expert panel will be headed by B Sesikeran. Mr. Sesikeran is the former Director of National Institute of Nutrition (NIN) and comprises Hemalatha and Dr. Nikhil Tandon.

1st All India Rural Financial Inclusion Survey by NABARD Released

i) *NABARD All India Financial Inclusion Survey (NAFIS),* conducted by National Bank for Agriculture and Rural Development (NABARD), was released by Dr. Rajiv Kumar, Vice Chairman, NITI Aayog in New Delhi. The report revealed that farm households register higher income than the families solely dependent on non-farm livelihood activities in rural areas.

ii) The survey, with reference year of 2015-16, which covered **40,327 rural households,** highlighted that the average annual income of an agricultural household is Rs. 1,07,172 compared to Rs. 87,228 for families engaged only in non-agricultural activities.

India's 1st Biofuel-Powered Flight Lands in Delhi

i) In a Historic day for the country's aviation and energy sector, India's first biofuel-powered flight landed in Delhi, covering the distance between Dehradun & Delhi. The bio-fuel is developed by CSIR-Indian Institute of Petroleum (IIP), Dehradun. The flight test was successfully completed by SpiceJet.

ii) The flight was powered with a blend of 75% air turbine fuel (ATF) and 25% bio jet fuel. The advantage of using bio jet fuel as compared to ATF is that it reduces carbon emissions and enhances fuel efficiency.

ICAR Organizes 2-Day Conference on Motivating & Attracting Youth in Agriculture

i) In order to meet the challenge of providing sustainable livelihoods for a rapidly growing population and to motivate and attract youth in agriculture, Dr. Trilochan Mohapatra, Secretary (DARE) & Director General (ICAR), inaugurated a two-days conference at NASC, New Delhi on Motivating and Attracting Youth in Agriculture (MAYA).

ii) The two-days conference provides an opportunity to all stakeholders to interact and discuss various options and avenues for not only attracting youth to agriculture but also even motivating them towards entrepreneurship in agriculture and allied fields.

Agriculture Minister Chairs Inter-Session Meeting On 'Marine Fisheries-Mariculture in India'

i) The Agriculture and Farmers' Welfare Minister Radha Mohan Singh has notified the "National Policy on Marine Fisheries, 2017", which will guide the development of marine fisheries sector for the next 10 years.

ii) It was announced at the inter-session meeting of the consultative committee of the Ministry of Agriculture and Farmers' Welfare on ***"Marine Fisheries-Mariculture in India"*** in Rameshwaram, Tamil Nadu.

iii) The government has also included a sub-component, namely, 'Assistance for Deep-Sea Fishing' under 'Blue-Revolution'.

iv) The government has decided to sharply increase the Minimum Support Price (MSP) for Kharif crops, as it responds to concerns about distress in the country's farm economy despite two years of strong monsoons.

v) The MSP for common variety paddy, the most important Kharif crop, will be raised by Rs. 200 over last year's Rs. 1,550 per quintal. That works out a 13%t increase for the financial year ended March, 2019 compared to a 5.4% increase in FY-18. The year-on-year jump in MSP for most other Kharif crops has also been far steeper than the annual increases announced over the last three years.

NABARD Sanctions Over Rs. 735 crore for Bengal for Rural Infra Projects

i) NABARD has sanctioned **Rs. 735.53 crore** in the first quarter of the current fiscal under the Rural Infrastructure Development Fund (RIDF) for West Bengal for facilitating the execution of 86 projects.

ii) They include six solar power, one medium irrigation, five minor irrigation and 12 flood protection projects, besides 57 projects for the widening and strengthening of roads and 5 rural bridges.

Animals Declared As 'Legal Persons' in Uttarakhand

i) In a first, the Uttarakhand High Court accorded the status of ***'legal person or entity'*** to the animals in the state. The HC stated that "they have a distinct persona with corresponding rights, duties and liabilities of a living person."

ii) The unique status was bestowed upon the state's animals by a division bench of judges comprising Justice Rajiv Sharma and Justice Lokpal Singh while issuing a series of directions to prevent cruelty against animals.

Madhya Pradesh CM Inaugurated NTPC's Solar Plant

i) Madhya Pradesh Chief Minister Shivraj Singh Chauhan inaugurated state-run power giant National Thermal Power Corporation Limited (NTPC) 250 MW Solar Power Plant in Mandsaur.

ii) According to NTPC, this project has been set up mainly on non-agricultural land and generation of electricity has been achieved without affecting the natural eco-system with zero carbon emission at an investment of about **Rs. 1,500 crore**.

iii) This is one of the biggest solar plant developed using domestically manufactured solar cells and modules under the Make in India initiative.

PM Modi Dedicates Bansagar Canal Project to Nation in Mirzapur

i) The Prime Minister Narendra Modi dedicated the Bansagar Canal Project to the Nation in Mirzapur, Uttar Pradesh. This project will provide a big boost to irrigation in the region, and will be greatly beneficial for the farmers of Mirzapur and Allahabad districts of Uttar Pradesh.

ii) He inaugurated 100 Jan Aushadhi Kendras in the State. He also dedicated a bridge over the River Ganga at Balughat, Chunar, which will facilitate connectivity between Mirzapur and Varanasi.

SBI And NABARD Tie-Up for Credit in Telangana

i) State Bank of India and National Bank for Agriculture and Rural Development (NABARD) have joined hands to provide collateral-free credit through Joint Liability Groups (JLGs) in Telangana.

ii) The two signed an MoU, along with seven NGOs from Nizamabad, Medak and Warangal districts, for the promotion and credit linkage of 2,000 JLGs during the current year.

Odisha Government Launches 'Green Mahanadi Mission'

i) The Chief Minister of Odisha, Naveen Patnaik launched the ***'Green Mahanadi Mission'.*** The CM launched the mission by planting a sapling on the bank of river Mahanadi during his visit to Boudh and Subarnapur districts in western Odisha.

ii) The main objective of the mission is to stop soil erosion on river banks and recharge the groundwater reserve.

Karnataka Tops in Renewable Energy: IEEFA Report

i) According to the US-based Institute for Energy Economics and Financial Analysis (IEEFA) report, Karnataka is the new national leader in renewable energy generation. It has surpassed Tamil Nadu that had for quite some time been India's best renewable market.

ii) Karnataka has a sum of **12.3 gigawatts (GW)** of renewable capacity installed till March 2018. IEEFA's report was titled as "Karnataka's Electricity Sector Transformation".

Govt. Launches 'Krishi Kalyan Abhiyaan' to Help Farmers in Aspirational Districts to Raise Their Incomc

i) The government has announced the launch of 'Krishi Kalyan Abhiyaan' under which farmers in select villages will be given assistance and advice on ways to improve farming techniques and raise their income.

ii) The Agriculture Ministry in line with the Prime Minister Narendra Modi's vision of doubling farmers' income by 2022 has launched the Krishi Kalyan Abhiyaan from 1st June to 31st July, 2018.

Govt. Announces Rs. 8,000-Crore Package for Sugarcane Farmers

i) The Union government has announced a package of more than **Rs. 8,000 crore** aimed to provide relief to sugarcane farmers and to clear their dues. According to a government official, the package includes three elements. A buffer stock of 30 LMT (lakh metric tonnes) of sugar will be created with **Rs. 1,200 crore.**

ii) This money will be directly transferred to accounts of sugarcane farmers. **Rs. 4,400 crore** will be used to increase ethanol production to divert sugarcane away from sugar mills which are unable to dispose off their inventories. The government has also decided to ensure a minimum sale price of **Rs. 29/kg** for sugar.

Maharashtra Announces Financial Aid for Tur Farmers

i) Maharashtra government will provide the financial grant of **1,000 rupees per quintal** to those farmers whose tur and gram could not be purchased by the administration before the May 31 deadline.

ii) The state government had set up tur procurement target of **44.6 lakh quintals**. The government had started purchasing tur from farmers by offering them guaranteed rates following bumper tur production. Chief Minister Devendra Fadnavis stated the government will also give a financial aid of **160 crore rupees** to farmers, whose crops got damaged due to unseasonal rain or hailstorm. The Chief Minister also announced a special subsidy for milk producers.

Punjab Government Launched Mobile App- 'i-Hariyali'

i) The Punjab Government has launched a mobile application named **'i-Hariyali'** aimed at increasing the state's green cover. The mobile application would enable the users to order free plant saplings.

ii) The 'i-Hariyali' app, under 'Mission Tandarust Punjab' can be downloaded from the app store for free.

Gujarat Launches Wind-Solar Hybrid Power Policy

i) The Gujarat government announced wind-solar hybrid power policy - 2018, aiming to make optimum utilisation of the land and grid.

ii) The policy provides 50% concession for cross-subsidy surcharge and additional surcharge, when the generated energy is sold to a third party. Gujarat currently generates over **5,500 MW** of wind power and **1,600 MW** of solar power.

Gujarat Launches 'Suryashakti Kisan Yojna' for Farmers

i) The Gujarat government launched a solar power scheme for farmers- Suryashakti Kisan Yojna (SKY) enabling them to generate electricity for their captive consumption as well as sell the surplus power to the grid and earn an extra buck.

ii) As per the scheme, farmers having existing electricity connection will be given solar panels as per their load requirements. The State and Central governments will give 60% subsidy on the cost of the project. The farmer is required to take **5% cost**, while 35% will be provided to him as an affordable loan with interest rates of **4.5-6%.**

Maharashtra Government Forms Study Committee for Floating Solar Power Plant

i) Maharashtra government has set up a committee to study issues related to the development of a **1,000-MW** floating solar power plant at Ujani dam in Sholapur district. The committee will be headed by the Director (Commercial) of Maharashtra State Electricity Distribution Company Ltd. (MSEDCL), Satish Chavan.

ii) Among other things, the committee will study the environmental impact of the project and monitor the month-wise water level of the dam. Ujani is among the largest dams in Maharashtra.

Swachh Bharat Mission Launches GOBAR-DHAN Scheme

i) Union Minister for Drinking Water and Sanitation, Uma Bharti, launched the GOBAR (Galvanizing Organic Bio-Agro Resources) - DHAN scheme at the National Dairy Research Institute (NDRI) Auditorium, Karnal, Haryana.

ii) The scheme aims to positively impact village cleanliness and generate wealth and energy from cattle and organic waste.

NITI Aayog to Collaborate with IBM to Develop Precision Agriculture Using AI

i) NITI Aayog and IBM have signed a Statement of Intent (SoI) to develop a crop yield prediction model using Artificial

Intelligence to provide real-time advisory to farmers in Aspirational Districts. The Intent was signed in presence of NITI Aayog, CEO Amitabh Kant and IBM India, MD, Karan Bajwa.

ii) The first phase of the project will focus on developing the model for 10 Aspirational Districts across the States of Assam, Bihar, Jharkhand, Madhya Pradesh, Maharashtra, Rajasthan and Uttar Pradesh.

Telangana Government Launches Rythu Bandhu Scheme for Farmers

i) The Telangana government introduced *Rythu Bandhu,* a scheme to provide **Rs. 4,000 per acre** each to land-owning farmers for Rabi and Kharif seasons, and issued special passbooks for them.

ii) The programme was flagged off by Chief Minister K. Chandrashekhar Rao, who distributed 298 cheques to farmers at Huzoorabad Mandal in Karimnagar district. The state government has set aside **Rs. 12,000 crore** in the 2018-19 budget for Rythu Bandhu.

CM Pema Khandu Inaugurates Arunachal Pradesh's Biggest Solar Plant

i) Arunachal Pradesh Chief Minister Pema Khandu inaugurated the state's biggest solar power plant at Energy Awareness Park in the presence of Power Minister Tamiyo Taga in Itanagar.

ii) The plant, with a capacity of **1 MW**, was commissioned on April 1 at an estimated investment of **Rs. 8.50 crore**. The officials of **Arunachal Pradesh Energy Development Agency (APEDA)** successfully implemented the project.

Indian Tea Industry Records Highest Ever Production, Export in FY, 18

i) Indian tea industry has recorded the highest ever production as well as exports in FY, 18. The total tea production was **1325.05 million kgs**, an increase of **74.56 million kgs** as compared to 2016-17. In percentage terms, the increase is around **6%**.

ii) Similarly, the total quantity of tea exported during the financial year, 2017-18 stood at **256.57 million kgs**, while the foreign exchange realized from exports of Indian tea was **$ 785.92 million.**

iii) The growth in exports was majorly driven by the following countries: **Egypt** (increased by 7.49 million kgs), **Iran** (increased by 6.95 million kgs), **Pakistan** (increased by 4.96 million kgs), **China** (increased by 2.91 million kgs) and **Russia** (increased by 2.89 million kgs).

India Unveils National Wind-Solar Hybrid Policy

i) In yet another effort to boost clean energy generation, the Ministry of New and Renewable Energy (MNRE) unveiled a *National Wind-Solar Hybrid Policy* which seeks to promote new projects as well as hybridisation of the existing ones.

ii) Under the scheme, the government would bid out wind-solar hybrid projects on tariff-based competitive bidding and these projects can be set up anywhere across the country depending on feasibility and land availability with bidders.

India's 1st Solar Powered Railway Station in Guwahati

i) Guwahati now has India's first railway station run by solar power. The project of installing solar panels was commissioned in April, 2017. Around **2352 solar modules,** with a capacity of generating **700 kw** have been set up over the roof of the Guwahati railway station.

ii) Solar panels have been installed at the roof-top solar power plant. The solar-powered station is aimed at reducing carbon-footprint as well as cut down power costs drastically. The project cost is estimated to be **Rs. 6.7 crore,** as per Indian Railways.

Hamburg To Host 2018 Global Wind Summit

i) The first edition of the Global Wind Summit will be held in Hamburg, Germany in September, 2018. The four-days event will see speakers from about **100 countries**, including India, China, the US, Spain and Denmark.

ii) The conference on wind is the largest and most important meeting of the wind industry worldwide. The event combines two conferences, Wind Energy Hamburg and Wind Europe.

Cochin International Airport, World's 1st Fully Solar Energy-Powered Airport: UNEP

i) The United Nations Environment Programme (UNEP) has recognised Cochin International Airport Ltd. (CIAL) in Kerala as the world's first fully solar energy-powered airport. CIAL is India's first airport to be constructed in public-private-partnership (PPP) model in 1999.

ii) It is the busiest and largest airport in the state of Kerala. The airport has **15 MW** solar power plant near the cargo complex of airport comprising **46,150 solar panels** laid across **45 acres**.

Madhya Pradesh's Kadaknath Chicken gets Geographical Indication Tag

i) Six years after it applied for the Geographical Indications (GI) Tag for Kadaknath, a chicken breeds whose black meat is in demand in some quarters, Madhya Pradesh has got the coveted label.

ii) Chennai-based GI Registry accepted the application of Gramin Vikas Trust, Jhabua, which was filed in 2012.

Centre, Maharashtra Govt. & World Bank sign 420 Million Dollar Project

i) The Centre, Maharashtra Government and the World Bank have signed a 420-million-dollar project to help small and marginals farmers in the Marathwada and Vidarbha region of the state.

ii) The Maharashtra Project for Climate Resilient Agriculture will be implemented in rural areas largely dependent upon rain-fed agriculture. The project will take up a series of activities at the farm and watershed level.

World's Largest Solar Park to Come Up in Gujarat

i) Gujarat Chief Minister Vijay Rupani approved setting up a **5,000 MW** capacity solar park at the Dholera Special Investment Region (SIR), which would be the largest such entity in the world after its completion.

ii) The proposed solar power generation project would be set up in **11,000 hectares** of land with an investment of **Rs. 25,000 crore**.

Government Raises NABARD's Authorised Capital to Rs. 300 billion

i) As part of the drive to double the income of farmers by 2022, the government has raised the authorised capital of NABARD by six times from 50 billion to Rs. 300 billion to help it increase its lending to the rural economy.

ii) The increase in authorised capital will enable NABARD to respond to commitments it has undertaken, particularly in respect of the Long-Term Irrigation Fund and on-lending to co-operative banks.

NABARD Approves Rs. 14,690 cr Funding to Rajasthan

i) National Bank for Agriculture and Rural Development (NABARD) has extended total credit support of **Rs. 14,690 crore** to Rajasthan for 2017-18.

ii) Rajasthan has been one of the biggest beneficiaries under the Rural Infrastructure Development Fund (RIDF) of NABARD, under which NABARD disbursed concessional loan of **Rs. 1,851.29 crore** to the state government during 2017-18.

Diu, 1st to Run on 100% Renewable Energy During Daytime

i) Diu Smart City has become the first city in India, that runs on 100% renewable energy during daytime setting a new benchmark for other cities to become clean and green. Diu had been importing 73% of its power from Gujarat until 2017.

ii) To further enhance its solar capacity, Diu offers its residents a subsidy of Rs. 10,000-50,000 for installing **1-5KW** rooftop solar panels.

Rs. 10,012 cr Aid Given to UP in 2017-18: NABARD

i) According to the Development bank NABARD, it has provided an assistance of **Rs. 10,012 crore** to Uttar Pradesh during 2017-18 to facilitate overall development with special focus on agricultural credit.

ii) This has been achieved through commercial banks, RRBs and co-operatives, infrastructure development through dedicated funds, strengthening of rural financial institutions, women empowerment through Self Help Groups, micro-finance and several other new initiatives.

World's Largest Solar Park 'Shakti Sthala' Launched in Karnataka

i) The Karnataka government inaugurated the first phase of a **2,000 MegaWatts (MW)** solar park in the drought-prone district, Tumkur. The first phase of the **Rs. 16,500 crore** park called "Shakti Sthala" will generate **600MW**, while the balance **1,400MW** is expected to be commissioned by the end of 2018.

ii) The solar project is the largest in the world, is spread over **13,000 acres** and five villages. It is part of the "Karnataka Solar Policy 2014-2021" which aims to decrease dependence on traditional power sources and move to environmentally friendly ones to meet the growing power needs of the state. The park ties in with the centre's scheme to generate **100 GigaWatts (GW)** of solar power by 2020.

Milk Production Up 20% in 2014-2017

i) According to Minister of Agriculture and Farmers' Welfare Radha Mohan Singh, Milk production in the country increased by 20 per cent from **137.7 million tonnes to 165.4 million tonnes** between 2014 and 2017. Also, the per capita availability of milk has risen by **15.6 per cent from 307 gms a day to 355 gms per day** between 2013-14 and 2016-17.

ii) With a corpus of **Rs. 2,450 crore**, the government has set up an AHIDF for financing infrastructure requirement of animal husbandry sector.

CCEA Approves Continuation of Ongoing Urea Subsidy Scheme

i) The Cabinet Committee on Economic Affairs has approved the proposal of Department of Fertilizers to continue Urea Subsidy Scheme upto 2019-20 at a total estimated cost of **Rs. 1,64,935 crore** and for disbursement of fertilizer subsidy. This decision implies that there will be no increase in price of urea, till 2020.

ii) Urea Subsidy is a part of Central Sector Scheme of Department of Fertilizers with effect from 1st April, 2017 and is wholly financed by the Government of India through Budgetary Support.

KUSUM Scheme to Install Grid Connected Solar Plants in Rural India

i) The Indian government is in process of formulating 'Kisan Urja Suraksha evam Utthan Mahaabhiyan (KUSUM) scheme' to harness solar power for rural India. The scheme will allow to setup grid-connected solar plants in rural areas.

ii) KUSUM will build **10,000 MW** solar plant and provide **1.75 million** off-grid agricultural solar pumps. The program aims to boost farmers' income as they can sell their additional power to the main grid produced by solar plants located on their lands.

Union Agriculture Minister Addresses 5th India Maize Summit, 2018

i) Union Agriculture Minister, Radha Mohan Singh had addressed the 5th India Maize Summit in New Delhi.

ii) The aim of the summit was to harness the potential of maize, technological innovations, promoting producer aggregation and linkages that are required.

Union Agriculture Minister Addresses 5th India Maize Summit, 2018

i) Union Agriculture Minister, Radha Mohan Singh had addressed the **5th** India Maize Summit in New Delhi.

ii) The aim of the summit was to harness the potential of maize, technological innovations, promoting producer aggregation and linkages that are required.

Union Agriculture Minister Addresses 5th India Maize Summit, 2018

i) Union Agriculture Minister, Radha Mohan Singh had addressed the 5th India Maize Summit in New Delhi.

ii) The aim of the summit was to harness the potential of maize, technological innovations, promoting producer aggregation and linkages that are required.

NABARD Launches Rs. 1,918 crore Area Development Plan for Punjab

i) The National Bank for Agriculture and Rural Development (NABARD) announced area development plans worth **Rs. 1,918 crore** for Punjab to supplement farmers' income.

ii) Under this plan, the activities like dairy, vegetable cultivation, bee keeping, mushroom growing will be promoted in order to provide income to farmers at the sustainable basis.

Agriculture Ministry Adds 6 New Features in e-NAM Platform

i) Agriculture Minister Radha Mohan Singh launched six new features in the National Agriculture Market (e-NAM) platform to make it more user-friendly.

ii) The six new features of e-NAM platform includes:

1. MIS dashboard for better analysis,
2. BHIM payment facility by traders,
3. Mobile payment facility by traders,
4. Enhanced features on mobile app, such as gate entry and payment through mobile,
5. Integration of farmer's database, and
6. e-Learning module.

NER's 1st Regional Centre for Agriculture with Israeli Collaboration to be Inaugurated in Mizoram

i) Israeli Ambassador to India, Mr Daniel Carmon has stated that a Centre for Agriculture, with the collaboration of Israeli expertise, is going to be inaugurated in Mizoram in March, 2018.

ii) This will be the first such centre in the North-East region (NER) of India that is being established with Israeli collaboration. It has been set up at a cost of **Rs. 8-10 crore**.

NABARD Sanctioned Rs. 372.51 crore to Odisha

i) The NABARD has sanctioned **Rs. 372.51 crore** loan assistance under the Rural Infrastructure Development Fund (RIDF) to

Odisha towards irrigation and rural bridge projects for the last quarter of this fiscal.

ii) NABARD has also sanctioned **14 Rural Bridge Projects** under Biju Setu Yojna with RIDF loan assistance of **Rs. 242.34 crore**.

Andhra Pradesh Gramin Vikas Bank Introduces Desktop ATMs in Rural India

i) The Andhra Pradesh Gramin Vikas Bank (APGVB) has operationalised its first desktop ATM in Telangana. The mini-ATM within the branch premises will facilitate customers to withdraw small amounts.

ii) With these ATMs, there will be no need for separate ATM rooms, so there is no security issue. This initiative is the biggest under the SBI and among the best regional rural banks (RRBs).

Promotion of Zero Budget Natural Farming

- Indian Council of Agricultural Research (ICAR) under Network Project on Organic Farming (NPOF) & All India Co-ordinated Research Projects (AICRP) on Integrated Farming Systems, has initiated experiment on "Evaluation of zero budget farming practices in basmati rice-wheat system" at Modipuram (UP), Ludhiana (Punjab), Pantnagar (U'Khand) & Kurukshetra (Haryana) from rabi, 2017 to study zero budget farming practices on productivity, economics & soil health including soil organic carbon & soil fertility.
- Schemes for Zero Budget Natural Farming, Zero Budget Natural Farming, as name implies, is method of farming where cost of growing & harvesting plants is zero.
- This means that farmers need not purchase fertilizers & pesticides in order to ensure healthy growth of crops.
- It is natural farming technique that uses biological pesticides instead of chemical-based fertilizers.
- Farmers use earthworms, cow dung, urine, plants, human excreta & such biological fertilizers for crop protection.
- It reduces farmers' investment. It protects soil from degradation.

SOME UNIQUE QUALITY OF ZBNF

- In the Zero Budget Natural Farming **nothing** has to be purchased from the outside. All things required for the growth of the plant are available around the root zone of the plants.
- 98 to 98.5% nutrients are taken from air, water & solar energy.
- Remaining **1.5% nutrients** taken from the soil are also available **free** of cost as it is taken from the prosperous soil which is enriched with these nutrients.

Union Ministry Launched Meghalaya Milk Mission in Shillong

- Union Minister for Agriculture & Farmers Welfare Radha Mohan Singh launched Rs. 215 cr. Meghalaya Milk Mission in Shillong through National Co-operative Development Corporation (NCDC).

Objectives of Meghalaya Milk Mission

- In assisting to achieve Centre's goal of doubling farmer's income by 2022 through promotion of milk business.
- Amount will be utilised for training under dairy farm, chilling centre & purchase of milch animal.
- Project consists of **2000 dairy farm units.**
- Each unit houses 5 cows, shed constructions, storage room construction, animal insurance, 79 bulk milk coolers of 500 litre capacity, 13 milk tankers of **3000 litre capacity,** etc.
- **Rs. 1 cr.** would be additionally provided for training of farmers.
- Mission Sahakar-22 under NCDC is started to give employment opportunities through co-operatives.
- Creation of dedicated Meghalaya Dairy Development Agency is underway.
- Co-operative Development Project (ICDP) for computerization of primary co-operatives will be started. govt. will aid State co-operatives in areas of handicraft, handloom, organic farming, forest produce, horticulture, etc.
- Model Agriculture Produce and Livestock Contract Farming and Services Act, 2018.

Union Agriculture Minister released final Model Act titled 'The State/UT Agricultural Produce & Livestock Contract Farming & Services (Promotion & Facilitation) Act, 2018' on 22nd May, 2018. Main motive behind drafting Model Contract Farming Act is to integrate farmers w/bulk purchasers for better price realization.

Friends of Coconut Tree Program: India Makes Unprecedented Progress in Coconut

- India brought 13, 117 hectare (ha) under new coconut plantation b/w mid-2014 & 2018, vis-a-vis 9, 561 ha b/w 2010 & 2014. Productivity increased to 11, 516 fruits per ha in 2017 - 18 vis-a-vis 10, 122 in 2013 - 14.

Friends of Coconut Tree Program

- Under Friends of Coconut Tree, skill development programme for coconut sector, 33, 228 unemployed youths are trained as compared to 27, 770 b/w 2004 & 2014.
- The Coconut Development Board gave image makeover for coconut plucking by launching the "Friends of Coconut Tree" programme to train unemployed youth in the art of climbing coconut trees and caring for them.
- This was to alleviate acute shortage of manpower for plucking coconuts and caring for the trees.

Aims of Friends of Coconut Tree Program

- It aims to train about 5, 000 unemployed youth in this specialised skill and create a pool of trained personnel across the 10 coconut-growing districts of Kerala.
- Training will be in tree climbing, coconut harvesting, crown clearing and pest control operations.
- Pollination and hybridisation techniques, plant protection measures and identification of tender, mature and seed coconuts will also be taught.

Umbrella Scheme 'Green Revolution — Krishonnati Yojna' in Agriculture Sector.

- Govt. approved continuation of agriculture sector umbrella programme, 'Green Revolution–Krishonnati Yojna', w/central outlay of **Rs. 33, 269.97 cr**. till March, 2020. Programme, comprises of 11 schemes to develop agriculture & allied sector in holistic & scientific manner to increase income of farmers by enhancing production, productivity & better returns on produce.

- Cabinet Committee on Economic Affairs (CCEA) approved scheme as part of its objective to double farmers' income by 2022. Schemes will be continued w/expenditure of **Rs. 33, 269.976 cr.** for 3 financial years, i. e., 2017 - 18, 2018 - 19 & 2019 - 20 Schemes focus on creating & strengthening of infrastructure of production, reducing production cost & marketing of agriculture & allied produce.
- Schemes are under implementation for varying duration during past few years. In 2017 - 18, it was decided to club all these schemes under one umbrella.
- Schemes Part of Green Revolution – Krishnonnati Yojna:

Mission for Integrated Development of Horticulture (MIDH)

- Aims to promote holistic growth of horticulture sector & enhance horticulture production, improve nutritional security & income support to farm Households. It has total central share of **Rs. 7533.04 cr**.

National Food Security Mission (NFSM)

- It aims to increase production of rice, wheat, pulses, coarse cereals & commercial crops, through area expansion & productivity enhancement in suitable manner in identified districts of country. It will restoring soil fertility & productivity at individual farm level & enhancing farm level economy. It includes National Mission on Oil Seeds and Oil Palm (NMOOP), w/total central share of **Rs. 6893.38 cr**. It aims to augment availability of vegetable oils & to reduce import of edible oils.

National Mission for Sustainable Agriculture (NMSA)

- It aims at promoting sustainable agriculture practices best suitable to specific agro-ecology focusing on integrated farming, appropriate soil health management & synergizing resource conservation technology. It has total central share of **Rs. 3980.82 cr**.

Submission on Agriculture Extension (SMAE)

- It has total central share of **Rs. 2961.26 cr.** It aims to strengthen ongoing extension mechanism of State Govts, local bodies, etc., achieving food & nutritional security & socio-economic empowerment of farmers.
- It seeks to institutionalize programme planning & implementation mechanism, forge effective linkages & synergy amongst various stake-holders, support HRD interventions, promote pervasive & innovative use of electronic/print media, inter-personal communication and ICT tools, etc.

Sub-Mission on Seeds and Planting Material (SMSP)

- It has total central share of **Rs. 920.6 cr**. It aims to increase production of certified/quality seed, to increase SRR, to upgrade quality of farm saved seeds. Seed multiplication chain will be strengthen & new technologies and methodologies in seed production, processing, testing, etc. will promoted to strengthen and modernizing infrastructure for seed production, storage, certification & quality, etc.

Sub-Mission on Agricultural Mechanisation (SMAM)

- It has total central share of **Rs. 3250 cr**. It aims to increase reach of farm mechanization to small & marginal farmers & to regions where availability of farm power is low. Custom Hiring Centres to offset adverse economies of scale arising due to small landholding & high cost of individual ownership will, promoted. Hubs for hi-tech & high value farm equipment, create awareness among stakeholders through demonstration & capacity building activities will created under it.

Sub-Mission on Plant Protection and Plan Quarantine (SMPPQ)

- It has total central share of **Rs. 1022.67 cr**. It aims to minimize loss to quality & yield of agricultural crops from ravages of insect pests, diseases, weeds, nematodes, rodents, etc. & shield agricultural bio-security from incursions & spread of alien species. It will facilitate exports of Indian agricultural commodities to global markets, and promote good agricultural practices, particularly w/respect to plant protection strategies & strategies.

Integrated Scheme on Agriculture Census, Economics and Statistics (ISACES)

- It has total central share of **Rs. 730.58 cr**. It aims to undertake agriculture census, study of cost of cultivation of principal crops, to undertake research studies on agro-economic problems of country.

Integrated Scheme on Agricultural Co-operation (ISAC)

- It has total central share of **Rs. 1902.636 cr**. It aims to provide financial assistance for improving economic conditions of co-operatives, remove regional imbalances and to speed up co-operative development in agricultural marketing, processing, storage, computerization & weaker section programmes. It will help cotton growers fetch remunerative price for their produce through value addition besides ensuring supply of quality yarn at reasonable rates to decentralized weavers.

Integrated Scheme on Agricultural Marketing (ISAM)

- It has total central share of **Rs. 3863.93 cr**. It aims to develop agricultural marketing infrastructure & promote innovative & latest technologies & competitive alternatives in agriculture marketing infrastructure.
- It seeks to provide infrastructure facilities for grading, standardization and quality certification of agricultural produce. Nationwide marketing information network will establish to integrate markets through common online market platform to facilitate pan-India trade in agricultural commodities, etc.

National e-Governance Plan (NeGP-A)

- It has total central share of **Rs. 211.06 cr**. It aims to bring farmer centricity & service orientation to the programmes. Reach & impact of extension services will be enhanced & access of farmers to information & services throughout crop-cycle will be improved. It will enhance & integrate existing ICT initiatives of Centre & States & enhance.

Restructured National Bamboo Mission

- Cabinet Committee on Economic Affairs (CCEA) approved centrally sponsored scheme 'National Bamboo Mission' (NBM) under National Mission for Sustainable Agriculture (NMSA) on 25th April, 2018 for remaining period of 14th Finance Commission (2018-19 & 2019-20).

- Govt. has approved outlay of Rs. 1290 cr. for implementation of Mission during remaining period of 14th Finance Commission. Of total share, Rs. 950 cr. will be borne by centre.

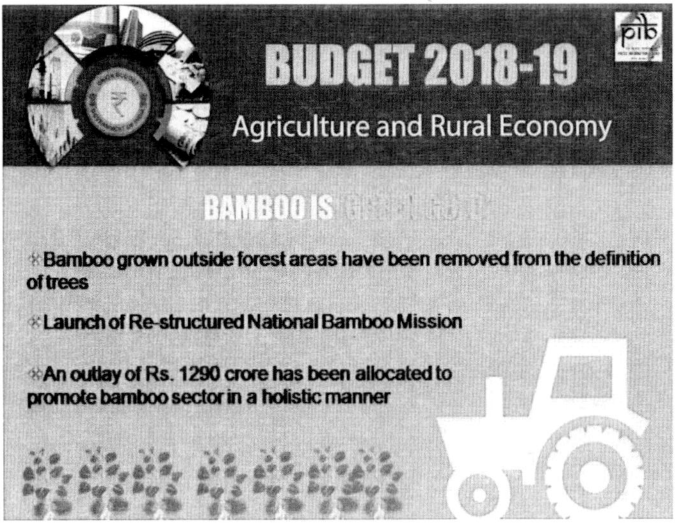

Restructured NBM Strives

- It will increase area under bamboo plantation in non-forest govt. & private lands to supplement farm income.
- It will improve post-harvest management through establishment of innovative primary processing units, treatment & seasoning plants, seasoning plants & market infrastructure.
- It will promote product development at micro, small & medium levels industry.
- It will promote skill development, capacity building & awareness generation for development of bamboo sector.

Integrated Scheme for Development of Silk Industry

- Sericulture Sector Central Sector Scheme "Integrated Scheme for Development of Silk Industry" is approved for next 3 years from 2017-18 to 2019-20.

- Scheme has 4 components: Research & Development (R&D), Training, Transfer of Technology and IT Initiatives, Seed Organizations and Farmers Extension Centres, Coordination and Market Development for seed, yarn and silk products, Quality Certification System (QCS) by creating chain of Silk Testing facilities, Farm based & Post-Cocoon Technology Up-gradation, and Export Brand Promotion.

Target of the Scheme

- Increase in production of bivoltine silk to around **8500 M < Ts p.a.** until 2020.
- Efficient Use of R&D for improvement in production of silk per hectare from **100 kg/hectare** to **110 kg/hectare**. Central govt. will provide improved reeling machines for Mulberry & Vanya Silk.
- Scheme aims to achieve self-sufficiency in silk production by 2022. To achieve this, production of high-grade silk in India will reach **20, 650 MTs** by 2022 from the current level of **11, 326 MTs**.
- There is clear focus on improving production of highest-grade quality of silk. It is proposed to increase 4A grade silk from the current level of **15 % to 25 %** of mulberry production by 2020.

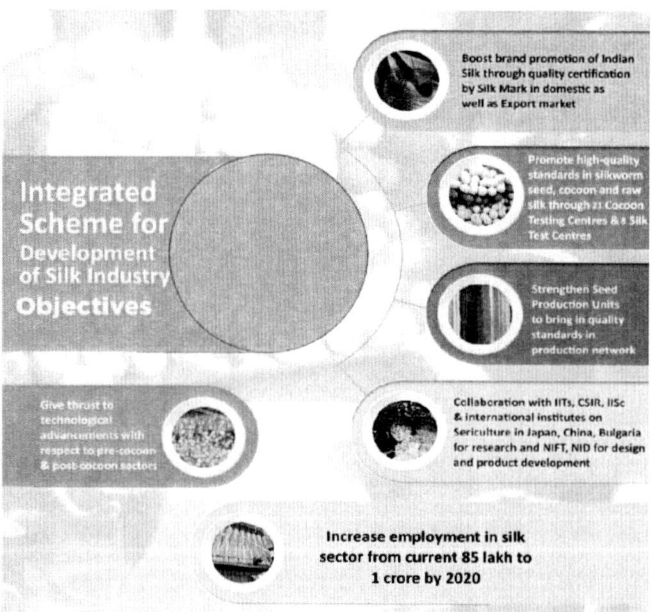

Rajasthan'S 1st Mega Food Park in Ajmer

- 1st Greentech Mega Food Park will be inaugurated at Roopangarh village in Ajmer, Rajasthan. Park is being set up at coast of Rs. 113.57 cr. & will benefit around 25,000 farmers in Ajmer Food Park is set up in area of **85.44 acres**. It would have Central Processing Centre & Primary Processing Centres (PPCs) at Jaipur, Nagaur, Tonk & Churu.

Inauguration of Uttarakhand'S Second Mega Food Park

- The Food Park is being set up in Uttarakhand at cost of **Rs. 99.96 cr**. & it gives benefit nearly **25,000 farmers**. Himalayan Mega Food Park Pvt. Ltd. is being set up in **50.14 acres area**. Park would have facilities for primary processing & storage near farms.

❑❑❑

Chapter 5

Agricultural Current Researches

Agricultural Current Researches

Pradhan Mantri Krishi Sinchayi Yojna: Cabinet approves Micro Irrigation Fund under NABARD

- The Cabinet Committee on Economic Affairs (CCEA) approved initial corpus of Rs.5,000 crore for setting up of dedicated Micro Irrigation Fund (MIF) in NABARD (National Bank for Agriculture and Rural Development) under Pradhan Mantri Krishi Sinchayi Yojna.
- The allocation of MIF under PMKSY will be made in instalments of Rs. 2,000 crore and Rs. 3,000 crore in 2018-19 and 2019-20. Under it, NABARD will extend loans to state governments during this period.
- The dedicated MIF will supplement efforts of Per Drop More Crop (PDMC) Component of PMKSY in an effective and timely manner. It will help to bring about 10 lakh ha of land under irrigation through innovative composite, community and cluster based micro irrigation projects with additional investment in MIF.

Green Revolution-Krishonnati Yojna: CCEA Approves Continuation of Umbrella Scheme in Agriculture Sector

- Cabinet Committee on Economic Affairs (CCEA) has approved continuation of Umbrella Scheme, Green Revolution - Krishonnati Yojna in agriculture sector beyond 12th Five Year Plan from 2017-18 to 2019-20. The Umbrella scheme comprises of 11 Schemes and Missions.
- They aim to develop agriculture and allied sector in holistic and scientific manner to increase income of farmers by enhancing production, productivity and better returns on produce. They

primarily focus on creating and strengthening of infrastructure of production, reducing production cost and marketing of agriculture and allied produce. These schemes will be now continued for three financial years, i.e., 2017-18, 2018-19 and 2019-20 with an expenditure of Rs.33,269 crore.

Government Launches Seed Import-Export Portal

- The Union Ministry of Agriculture and Farmers Welfare has launched seed import-export portal to facilitate export-import of seeds and planting material. It has been developed for making trade process easy. The ministry has removed all prohibitions imposed on export of seeds of different varieties to encourage export of seeds. It also has made import-export policy flexible and transparent for ascertaining availability of high-quality seeds to farmers found out everywhere in the world. The import-export of seeds and planting material is controlled under Import-Export Policy, 2002-2007.

CCEA approves restructured National Bamboo Mission

- The Cabinet Committee on Economic Affairs (CCEA) has approved restructured National Bamboo Mission (NBM), a Centrally Sponsored Scheme under National Mission for Sustainable Agriculture (NMSA) during remaining period of 14th Finance Commission (2018-20) with an outlay of Rs. 1290 crore. The restructured mission will ensure holistic development of bamboo sector by addressing complete value chain and establish effective linkage of producers (farmers) with industry.

Government Inks Agreement with World Bank for Maharashtra Project for Climate Resilient Agriculture

- Central Government, Maharashtra Government have signed US $420 million loan agreement with World Bank for Project for Climate Resilient Agriculture to help small and marginal farmers in Marathwada and Vidarbha regions of Maharashtra. The loan will be provided from World Bank & International Bank for Reconstruction and Development (IBRD). It will have maturity of 24 years and grace period of 6 years.

- The project aims to help in increasing climate resilient practices in agriculture and ensuring that farming continues to remain a financially viable activity for them. It will be implemented in rural areas largely dependent upon rainfed agriculture. It will take up series of activities at farm and watershed level.
- It will help to scale up climate-resilient technologies, such as micro irrigation systems, expand surface water storage and facilitate aquifer recharge, directly contributing to more efficient use of scarce water resources. It will help reduce risks of climate-related crop failure and enhance farmer's income by adopting climate-resilient seed varieties that are drought, salt tolerant and heat resistant.
- It will help to mainstream climate resilient agriculture agenda in various local institutions that deliver agricultural services to farming community. It will also help to improve capacity of farmer producers' organizations to operate as sustainable, market-oriented enterprises.
- This project is expected to benefit over 7 million people that are spread over area of 3.0 million hectares and cover 5142 villages across 15 most climate vulnerable districts of Maharashtra. It will help to mitigate emission of Green House Gases (GHGs) through carbon sequestration which involves planting trees in upper catchment areas and afforestation. It will also support Maharashtra government to shift towards new paradigm that puts climate resilience at core of agricultural growth and rural development in the state.

Government Inks Agreement with World Bank for Maharashtra Project for Climate Resilient Agriculture

- Central Government, Maharashtra Government have signed US $420 million loan agreement with World Bank for Project for Climate Resilient Agriculture to help small and marginal farmers in Marathwada and Vidarbha regions of Maharashtra. The loan will be provided from World Bank & International Bank for Reconstruction and Development (IBRD). It will have maturity of 24 years and grace period of 6 years.
- The project aims to help in increasing climate resilient practices in agriculture and ensuring that farming continues to remain a

financially viable activity for them. It will be implemented in rural areas largely dependent upon rainfed agriculture. It will take up series of activities at farm and watershed level.

- It will help to scale up climate-resilient technologies, such as micro irrigation systems, expand surface water storage and facilitate aquifer recharge, directly contributing to more efficient use of scarce water resources. It will help reduce risks of climate-related crop failure and enhance farmer's income by adopting climate-resilient seed varieties that are drought, salt tolerant and heat resistant.
- It will help to mainstream climate resilient agriculture agenda in various local institutions that deliver agricultural services to farming community. It will also help to improve capacity of farmer producers' organizations to operate as sustainable, market-oriented enterprises.
- This project is expected to benefit over 7 million people that are spread over area of 3.0 million hectares and cover 5142 villages across 15 most climate vulnerable districts of Maharashtra. It will help to mitigate emission of Green House Gases (GHGs) through carbon sequestration which involves planting trees in upper catchment areas and afforestation. It will also support Maharashtra government to shift towards new paradigm that puts climate resilience at core of agricultural growth and rural development in the state.

Uttarakhand's Second Mega Food Park Inaugurated at Udham Singh Nagar

- Uttarakhand's second Mega Food Park- Himalayan Mega Food Park Private Ltd. inaugurated at Mahuakhera Ganj in Kashipur in Udham Singh Nagar district. It was inaugurated by Union Food Processing Industries Minister Harsimrat Kaur Badal. The first Mega Food Park in the state is located at Haridwar and has already become operational.
- The scheme aims to create modern infrastructure facilities for food processing along value chain from farm to market with strong forward and backward linkages through cluster-based approach. Projects under it are merged under comprehensive flagship scheme of 'Pradhan Mantri Kisan Sampada Yojna'.

Rajasthan's First Mega Food Park Inaugurated at Ajmer

- Rajasthan's first mega food park-Greentech Mega Food Park Private Ltd. was inaugurated at Roopangarh Village in Ajmer district. It was inaugurated by Union Minister for Food Processing Industry Harsimrat Kaur Badal. The park was set up under Central Government's flagship Pradhan Mantri Kisan Sampada Yojna and 'Make in India' initiative.

CCEA approves bulk export of all edible oils

- The Cabinet Committee on Economic Affairs (CCEA) has approved for removal of prohibition on export of all varieties of edible oils except mustard oil. The proposal was forwarded by Ministry of Commerce & amp; Industry. The CCEA has also approved empowering Committee, chaired by Secretary, Department of Food & amp; Public Distribution to review the export and import policy on all varieties of edible oils.
- The MoU will help to increase co-operation between both ministries and help in fulfilling Government's dream of "Kaushal Bharat-Kushal Bharat". It will help to increase agricultural productivity, post-harvest management and fair-prices to the farmers for their produce, lower risk in agriculture and strengthen and develop other aspects of farmers, income such as horticulture, animal husbandry, beekeeping, dairying, fisheries, etc.

Maharashtra's first Mega Food Park inaugurated at Satara

- Maharashtra's first Mega Food Park named as Satara Mega Food Park Pvt. Ltd. was inaugurated at Degaon Village in Satara district by Minister of Food Processing Industries Harsimrat Kaur Badal. This is overall 12th Mega Food Park operationalized in country so far and 10th operationalized during tenure of present government.

Government to launch GOBAR-Dhan Yojna

- The Union Government is going to roll out Galvanizing Organic Bio-Agro Resources Dhan (GOBAR-DHAN) scheme. It was first announced by Finance Minister Arun Jaitley during his 2018-19 budget speech.

- The scheme focuses on managing and converting cattle dung and solid wastes in farms to useful compost, biogas and bio-CNG. It will also help in keeping the village clean while increasing income of farmers and cattle herders.
- Significance of Scheme.
- The scheme will be hugely beneficial for country as India is home to highest cattle population in world (around 300 million in number) which produces around 3 million tonnes of dung. It will encourage farmers to consider dung and other wastes not just as waste but as source of income.

North-East India's First Regional Centre for Agriculture with Israel Collaboration to come up in Mizoram

- North-east India's first regional Centre for Agriculture with collaboration of Israel will be inaugurated in Mizoram in March, 2018. It will be first such centre in North-East region of India being established with Israeli collaboration.
- There are 22 such operational centres setup across India with support of Israel including in states of Haryana, Gujarat, Madhya Pradesh, Rajasthan and Punjab. The first centre was established in 2008, in Haryana
- It will benefit the farmers of the north-eastern region in the long run and will also encourage the farmers from other regions. It will serve as model of learning for other countries in Indian sub-continent. It will benefit the people of North-East in the area of innovation, especially for start-ups.

Government Launches Six New User-friendly Features of e-NAM Platform

- Union Ministry of Agriculture & amp; Farmers' Welfare have launched six new features of National Agriculture Market (e-NAM) Platform to make it more users friendly.
- The six new features are e-NAM Mobile App, BHIM payment facility, new and improved Website with e-Learning Module, MIS Dashboard, Grievance Redressal Management System for Mandi Secretaries and Integration with Farmer Database.

- e-NAM is one of the major and important flagship schemes implemented by Ministry of Agriculture & amp; Farmers' Welfare. Its objective is to provide competitive and remunerative price to farmers for their produce through online competitive and transparent bidding process.
- e-NAM is pan-India e-trading platform. It is designed to create unified national market for agricultural commodities. Its objective is to provide competitive and remunerative price to farmers for their produce through online competitive and transparent bidding process.

Punjab Government makes super straw management mandatory.

- Punjab government has made attachment of super straw management system to combine harvester machines mandatory for harvesting paddy (rice) crop in an attempt to check the dangerous trend of stubble burning. The instructions were issued under section 31 A of the Air (Prevention and Control of Pollution) Act, 1981.

CCEA approves hike in MSP for Copra for 2018 season

- The Cabinet Committee on Economic Affairs (CCEA) has approved increase in Minimum Support Price (MSP) for Fair Average Quality (FAQ) of Milling Copra for 2018 season. The approval was based on recommendations of Commission for Agricultural Costs and Prices (CACP).
- The increased in MSP of Copra is expected to ensure appropriate minimum prices to farmers and set up investment in Coconut cultivation and thereby production and productivity in country.

Punjab Government bans sale of 20 insecticides...

- Punjab Government's Department of Agriculture has issued directions to discontinue the sale of 20 pesticides (insecticides) including Endosulfan harmful to health of humans and environment with immediate effect. This decision was taken based on the recommendations of Registration Committee, Punjab Agricultural University (PAU) and Punjab State Farmers & Commission (PSFC).

- The insecticides that are banned includes Tricholorofon, Phosphamidon, Dicofol, Methomyl, Thiophanate Methyl, Benfuracarb, Bifenthrin, Endosulfan, Chlorfenapyr, Carbosulfan, Ethofenprox (Etofenprox), Phorate, Triazophos, Dazomet, Diflubenzuron, Fenitrothion, Metaldehyde, Kasugamycin, Alachior and Monocrotophos.

Himachal Pradesh launches Zero Budget Natural Farming Project to promote Organic Farming

- Himachal Pradesh Government has launched Zero Budget Natural Farming (ZBNF) project to promote organic farming. It was launched by Chief Minister, Jai Ram Thakur. The project aims to increase agriculture produce and the income of farmers by the year 2022.

Zero Budget Natural Farming (ZBNF)

- ZBNF is set of natural farming methods where cost of growing and harvesting plants is zero. It is a farming practice that believes in natural growth of crops without adding any fertilizers and pesticides or any other foreign elements. It is different from organic farming.
- The word Zero Budget refers to zero net cost of production of all crops. This means that farmers need not purchase fertilizers and pesticides in order to ensure the healthy growth of crops. The inputs used for seed treatments and other inocluations are locally available in the form of cow dung and cow urine.
- It requires almost no monetary investment and envisages use of 'Jeevamrutha' and 'Beejamrutha'. The main aim of ZBNF is eliminate use of chemical pesticides and uses biological pesticides and promote of good agronomic practices. Farmers use earthworms, cowdung, urine, plants, human excreta and such biological fertilizers for crop protection.
- It lowers cost of inputs of farmers and gives better capacity to increase the incomes. It also protects soil from degradation and helps in retaining soil fertility and is climate change resilient.

Maharashtra approves Nanaji Deshmukh Krishi Sanjivani Yojna to promote climate-resilient Agriculture

- **Maharashtra** Government has approved Nanaji Deshmukh Krishi Sanjivani Yojna, a Rs.4,000-crore project aimed at promoting climate-resilient agriculture. It will be roll out in 2018-19 and continue till 2023-24. The scheme will be implemented in 5,142 villages across 15 districts. Its objectives are to improve soil quality, develop foodgrain varieties which can sustain climate variations and effect necessary changes in the crop pattern as per the availability of water in a particular region.
- The scheme will cover small and medium- scale farmers, who are more vulnerable to the impact of climate change. The total cost of project is Rs. 4,000 crore, 70% of which will be borne by World Bank while state will contribute 30% over six years.

10th Global Forum for Food & Agriculture held Berlin...

- The 10th edition of Global Forum for Food & amp; Agriculture was held in Berlin, Germany from 18th -20th January, 2018. The theme of this year's event is "Shaping the future of livestock - sustainably, responsibly, efficiently.
- Global Forum for Food and Agriculture (GFFA).
- The GFFA is international conference that focuses on central questions concerning future of global agri-food industry. It is world's largest conference of agriculture ministers. It is held during International Green Week (IGW).
- The National Commodity and Derivatives Exchange Ltd. (NCDEX) unveiled India's first agricultural-commodity options in Guarseed designed as hedge for farmers to safeguard against price risks. It was launched by Union Finance Minister Arun Jaitley in New Delhi.

NCDEX launches India's first agri-options trading in Guarseed

- NCDEX's guarseed option is one of the most liquid contracts on the exchange platform. Its three options contracts in guarseed will expire in February, March and April, 2018 and will be made available for trading.

- NCDEX is the second exchange after MCX to launch options trading in commodities. In October, 2017, MCX had launched gold options. Market regulator SEBI had allowed NCDEX to launch options in September, 2016 and later had clarified that options will have futures as underlying for settlement.

Fourth ASEAN-India Ministerial Meeting on Agriculture & Forestry held in Delhi

- The 4th ASEAN-INDIA Ministerial Meeting on Agriculture and Forestry was held at NASC Complex, New Delhi. It was attended by representatives from 10 ASEAN countries.
- The aim of the meeting was to facilitate promotion of joint research for development of technologies for increasing production and productivity of crops, livestock and fisheries, and natural resources management, among others in the region.
- It also aimed at development of joint ventures, exchange of technologies, expertise and material. The meeting also aimed to enhance co-operation on exchange of expertise to promote resilience in food security to address the price volatility of food supplies in the region.
- The meeting held discussions focused on enhancing co-operation on exchange of expertise to promote resilience in food security and to address the price volatility of food supplies in the region. It was announced that 5th ASEAN-India Ministerial Meeting on Agriculture and Forestry will be held in 2019, in Brunei Darussalam.

New Research on Agri Products

Geoportal 'BHOOMI'

- The NBSS & LUP developed NBSS BHOOMI Geo-portal to access various thematic information on major physiographic regions of India, sub-physiographic regions of India, agro-ecological regions (1992), agro-ecological regions (2015) and agro-ecological sub-regions of the country.
- The point database for benchmark soil information of India for 300 sites, statewise soil series of India for 18 states (3,795 sites), soil series database for 22 districts (495 sites), grid point data on

soil physical and chemical properties of 752 NATP (National Agricultural Technology Projects) sites and soil fertility data (20,7269 sites) were processed and uploaded on Geoportal.

Wealth from waste

- Lac mud is the waste product of lac processing industries which is about 2.5–4.5% on dry and wet weight basis, respectively, of the raw material (sticklac) processed.
- Application of decomposed enriched lac mud in vegetables produced 22.0, 22.5 and 18.3% higher yield of brinjal, tomato and spinach, respectively, over 100% N through inorganic source (farmers' practice). This technology saves 48% of N and P fertilizers, and 65% of K fertilizer in brinjal and tomato. In spinach, similar saving in N and P fertilizers along with 36.6% saving of K fertilizers was recorded.
- Application of fortified lac mud in floriculture produced 31.7 and 38.5% higher flower yield of rose and chrysanthemum, respectively, over conventional method of manuring, ensuring saving of 5 kg and 700 g of manure/plant in rose and chrysanthemum, respectively.

A novel method of insect assisted controlled pollination

- A novel technique was developed for insect facilitated controlled pollination in oil palm. It involves covering the female inflorescence with a pollination bag, securing it at the bottom that no insects or pollen enter from outside the pollination bag, cleaning the pollinating insects and storing in a container, and transferring insects along with pollen of oil palm to reach the bagged inflorescence through a tube.
- The weevils along with pollen could reach a height of 22 feet in 6 minutes under controlled conditions. Fruit setting is normal. The fruit ratio, seed germination and seedling establishment are on par with manual pollination.

Breeding buck production through multiplier flocks

- Multiplier flocks of Barbari goats were adopted to popularize scientific goat farming. Various interventions were provided to these flocks for development of livelihood model and agri-business models for breeding and goat production.

- These flocks were established in Uttar Pradesh, Haryana and Rajasthan. Overall survivability at multiplier flocks was 93.3%. These farms have mostly focussed on increasing flock size; and sale restricted to 15-40%. The cost of production per goat/year varied from 3,260 to 7,800 and profit from 4,300 to 9,600 with an average of 5,280.

AICRP on Goat Improvement

- Improving goat productivity in farmer's flock under different agro climatic conditions: The project is covering 13 registered breeds and 3 local genotypes (lesser-known goats) in different parts of the country.
- AICRP on Goat Improvement is operational at 464 villages covering 3,010 farmers. The performance recording was carried out in 25,622 animals during the year.
- The increase in body weight at 12 months age over the units varied from 0.9 to 6.2%. Similarly, the increase in milk yield at 90 days varied from 3.4 to 8.8% over the units. The average pashmina production of Changthangi goats was 269.11 g in Laddakh region.
- Preventive health care, provided to 97,546 animals in farmers' flock, resulting in lower mortality rates ranging from 2.78 to 11.15%, thus increasing population growth and improving farmers' income from 22 to 35%.
- The project covered 13 tribal villages and contributed for a better livelihood security for tribal people, as goats are major source of income to poor people in tribal areas and NEH region.

Guinea fowl-breaking seasonality in breeding behaviour

- The Guinea fowl is a seasonal breeder and breeds in hot season (Apr-Sept) due to photoperiod effect, thus limiting its production and reproduction. In an effort to break the seasonality and to achieve round the year production from guinea fowl, different combinations of dietary protein and photoperiod were evaluated during winter (Oct - Feb) in Pearl guinea fowl.
- Birds (23) fed with 20% dietary protein and reared under18 hr photo-period produced 533 (HHEP 25%) eggs during Nov-Feb

with a fertility of 53.5 and hatchability as 89%. This is for the first time that seasonality has been successfully broken in breeding behaviour of guinea fowls.

Minimizing pesticide residues in agricultural produce

- A total of 246 multi-location supervised field trials were conducted to evaluate persistence and dissipation of new and existing pesticides on various crops in different agro-climates. The data were utilized to fix maximum residue limit (MRL) for 10 pesticide crop combinations by the Food Safety Standards Authority of India (FSSAI).
- Their safe waiting period and label claims were approved by the Central Insecticide Board and Registration Committee. After monitoring 22,270 samples of food commodities and water pesticide residues were detected in 4,289 samples (19.3%), while 473 (2.1%) samples exceeded FSSAI MRL. Based on the data submitted to Joint Meeting of Pesticide Residues (JMPR), the Codex Committee on Pesticide Residue (CCPR) adopted the MRL of 19 pesticides on spices.

CIARI Bio-Consortia

- The talc-based bio-formulation with consortia of antagonistic bacteria was developed and tested in different field conditions for management of bacterial wilt in brinjal.
- Soil application of bio-enriched FYM with CIARI-Bioconsortia at 15 days regular intervals could successfully control the bacterial wilt disease and increase the yield of brinjal. Soil application of bio-enriched FYM was also considerably good in protected cultivation.

Indian Marine Fisheries Code

- The Code of Conduct for Responsible Fisheries (CCRF), 1995 sets out international principles and standards of behaviour to ensure effective conservation, management and development of living aquatic resources, marine as well as freshwater, considering both the impact of fishing on ecosystems and the impact of ecosystems on fisheries and the need to conserve bio-diversity.

- ICAR-CMFRI and ICAR-CIFT jointly developed guidelines to put in practice the FAO's Code of Conduct for Responsible Fisheries (FAOCCRF, 1995), and named it 'Indian Marine Fisheries Code (IMFC)' to give impetus to bring about changes in the manner in which marine fisheries is managed in the country.
- IMFC explains in detail each sub-article of the FAOCCRF and provides information on how the article can be implemented and by whom. IMFC further proposes several new bodies which are necessary to place fisheries management in the country on an unassailable footing.

Food fishes classified based on their fat content

Lean fish (<2% fat)	Low fat fish (2-4% fat)	Medium fat fish (4-8% fat)	High fat fish (> 8% fat)
Finfishes: *Euthynnus affinis, Ailia coila, Rastrelliger kanagurta, Rita rita, Katsuwonus pelamis, Stolephorus commersonii, Stolephorus waitei, Epinephelus* spp., *Sperata seenghala, Xenentodon cancila, Thunnus albacares* Shellfishes: *Perna viridis, Fenneropenaeus indicus, Penaeus monodon*	Finfishes *Leiognathus splendens, Oncorhynchus mykiss, Clarias magur, Trichiurus lepturus, Mugil cephalus, Neolissochilus hexagonolepsis, Cyprinus carpio, Catla catla, Cirrhinus mrigala, Heteropneustes fossilis, Labeo rohita, Lates calcarifer, Schizothorax richardsonii, Harpodon nehereus* Shellfishes: *Crassostrea madrasensis*	Finfishes: *Anabas testudineus, Gudusia chapra, Nemipterus japonicas, Puntius sophore, Etroplus suratensis, Amblypharyngodon mola, Tor putitora* Shellfishes: *Macrobrachium rosenbergii*	Finfishes: *Tenualosa ilisha, Sardinella longiceps*

Organic Animal Husbandry

- Organic certification of fodder was carried out and after due inspection and auditing, third year organic scope certificate was awarded for the fodder as per the NPOP standards of India. NRC on Meat developed the protocols for organic sheep production and recently received organic livestock certification from Organic India.

Ranking of Agricultural Universities

- Agricultural Universities Ranking was made to improve ranking of Indian universities in World University Rankings. The emphasis on research product, research impact, research excellence, technologies transferred to farmers, spread/adoption of technologies and increase in agricultural growth in the area of jurisdiction of the university, etc. were considered while evaluating the agricultural universities.

- Based on the evaluation, 57 agricultural universities have been ranked and the following top three universities (ICAR-National Dairy Research Institute, Karnal; ICAR Indian Agricultural Research Institute, New Delhi; Punjab Agricultural University, Ludhiana) were honoured on ICAR Foundation Day Award Ceremony held on 16 July, 2017.

Germplasm Registration Information System (GRIS)

- The Germplasm Registration Information System (GRIS) was developed to make the entire process of germplasm registration—submission of application, evaluation by experts and decision by Plant Germplasm Registration Committee—a web-based system.
- The GRIS was launched by DG, ICAR on 2 August, 2017. The system is expected to provide gene bank managers, breeders and plant researchers with a hands-on-tool for management of germplasm registration process, and to policy-makers with a reliable source of information. To register and use GRIS, one can log on to www.nbpgr.ernet.in/registration.

ICAR Data Centre

- Shri Radha Mohan Singh, Union Minister for Agriculture and Farmers Welfare, inaugurated the ICAR Data Centre and launched the KVK Mobile App at ICAR-Indian Agricultural Statistics Research Institute, New Delhi.
- ICAR Data Centre provides the ICT infrastructure for hosting the web applications developed by ICAR Institutes; the unified messaging solution which is based on Microsoft Exchange Server, which helps ICAR personnel to stay connected with each other via instant messaging, email, audio/video calls, persistent chat rooms, online meeting and presentations for scientific/research/technological/educational/extension information exchange; uniform email IDs (https://mail.icar.gov.in) of all ICAR personnel under single ICAR domain for effective communication. ICAR Data Center is ISO-27001:2013 certified for Information Security and ISO 20000-1:2011 for IT Service Management System.

Agricultural Research Data Book

- The Agricultural Research Data Book (ARDB) 2017 was published. The ARDB, 2017, which is 20th in the series, is an attempt to put together main components/indicators of such information and comprises 170 tables organized into 10 sections namely, Natural Resources, Agricultural Inputs, Animal Husbandry, Dairying and Fisheries, Horticulture, Production and Productivity, Agricultural Engineering and Produce Management, Export and Import, India's Position in World Agriculture, Investment in Agricultural Research and Education and Human Resources under National Agricultural Research System (NARS).
- Apart from updating of tables, 22 new tables added, which consist of state-wise data on area and production of 7 food grains rice, wheat, jowar, bajra, maize, pulses, oilseeds, and 5 fruit crops mango, banana, sweet orange, apple and papaya; 3 vegetable crops potato, tomato and onion; production and per capita availability of livestock products like milk, egg, and production of meat and wool. For depicting state-wise data, thematic maps were prepared using Geographical Information System (GIS).

Avadh Mango Grower Society

- This programme has made its mark in implementation areas through creation of SHG, Farmer Interest Group (FIG), Society and Farmer Producer Organization (FPO).
- For instance, Avadh Mango Grower Society, Lucknow was formed to promote production and marketing of organic Malihabadi Dusseri Mango and they have obtained Geographic Indicator (GI-125) with the help of Farmer First Project Centre at CISH, Lucknow, which helped the farmers to sell their mango at the premium rate of 60/kg.

Successful introduction of kiwi fruit in Wokha

- District Nagaland, The Wokha hill district of Nagaland has vast potential for the cultivation of high value kiwi fruit. Realizing its importance, kiwi cultivation was introduced at New Wokha Village in 2 ha of area.

- Grafted kiwi, namely Allison (female) and Tomuri (male) brought from NBPGR, Regional Station, Bhowali, Nainital (Uttarakhand) were planted by each of the selected farmer. Around 70 % of the saplings survived with good vegetative growth.
- During the year 2016, the first harvest was obtained with 5 - 8 kg Kiwi fruits per vine. Being a new crop, therefore, most of the harvested fruits were gifted by the farmers. Only little quantity was sold @ 150/ kg. As kiwi fruit has high value in the market and cost around 200-250/ kg, farmers may reap the benefit in forthcoming years.

Plant Genome Saviour Community Award

- Khola/Canacona Chilli Cultivators Group, South Goa was awarded with the prestigious Plant Genome Savior Community Award by PPV & FRA, New Delhi. The awards were conferred by Sri Radha Mohan Singh, Union Minister for Agriculture and Farmers Welfare, Government of India on 21 December, 2016, for conservation and preservation of Khola/ Canacona Chilli, a medium pungent land race with bright red colour.
- In the same function, Sri Radha Mohan Singh, Union Minister for Agriculture and Farmers Welfare, awarded Rev. Saw Saytha, President of Karen Welfare Association, Mayabunder, North and Middle Andaman.
- The community was honoured for its contribution to conservation and preservation of traditional and unique rice varieties, viz.Khushbuyya, Black Burma, White Burma, Red Burma, Mushley and Nyaw-in, etc., since 1925 in the islands.

Department of Agricultural Research and Education

- The Department of Agricultural Research and Education (DARE) was established in the Ministry of Agriculture, Government of India in December, 1973 to co-ordinate and promote agricultural research and education in the country.
- DARE provides the necessary government linkages for the Indian Council of Agricultural Research (ICAR), the premier research organization for co-ordinating, guiding and managing research in areas including crop science, horticultural science natural resource management, agricultural engineering, animal science,

fisheries science, agricultural education and agriculture extension in the entire country.

- With 112 ICAR Institutions and 74 Agricultural Universities spread across the country, this is one of the largest national agricultural research systems in the world. Apart from ICAR, the Department of Agricultural Research and Education has other autonomous bodies, viz. the Central Agricultural Universities (CAUs) at Imphal (Manipur), Jhansi (Uttar Pradesh), and Pusa (Bihar); and AgrInnovate India Limited, Delhi under its administrative control.

Know About - Indian Council of Agricultural Research

- The Indian Council of Agricultural Research is an autonomous organization under the Department of Agricultural Research and Education, Ministry of Agriculture and Farmers Welfare, Government of India. Formerly known as Imperial Council of Agricultural Research, it was established on 16 July, 1929 as a Registered Society under the Societies Registration Act, 1860 on the recommendations of the Royal Commission of Agriculture.
- It was reorganized in 1965 and again in 1973, with its Headquarters located in Krishi Bhawan, New Delhi, with support facilities in Krishi Anusandhan Bhawan I and II and NASC Complex, Pusa, New Delhi.
- The Union Minister of Agriculture and Farmers Welfare is the President of the ICAR. The Principal Executive Officer of the ICAR is the Director General, who also functions as Secretary, Department of Agricultural Research and Education, Government of India.
- The General Body of the ICAR Society, headed by the Union Minister of Agriculture and Farmers Welfare is the supreme authority of the ICAR.
- Its members include Ministers for Agriculture, Animal Husbandry and Fisheries, and the senior officers of the various state governments, Members of Parliament and the representatives from industry, research institutes, scientific organizations and farming community.
- The Governing Body headed by the Community Director General, who is also the Secretary, DARE is the chief executive and

decision-making authority of the ICAR. The Governing Body consists of eminent agricultural scientists, educationists, public representatives and representatives of the farmers.

- It is assisted by the Accreditation Board, Regional Committees, Policy and Planning Committee, several Scientific Panels and Publications Committee. In scientific matters, the Director General is assisted by eight Deputy Directors General, one each in (i) Crop Science, (ii) Horticultural Science, (iii) Natural Resource Management, (iv) Animal Science, (v) Agricultural Engineering, (vi) Fisheries Science, (vii) Agricultural Education, and (viii) Agricultural Extension, who are also assisted by Assistant Directors General, and are the Heads of their Subject Matter Divisions (SMDs) for the entire Country.
- SMDs are responsible for extending all technical and financial guidance and support to the Research Institutes, National Research Centres and the Project Directorates within their respective Divisions.
- In addition, Assistant Directors General National Agricultural Science Fund (NASF), Co-ordination, Plan Implementation and Monitoring, Intellectual Property and Technology Management, International Relations and Human Resources Management also assist the Director General in their respective job roles.
- The ICAR also recruits scientists and other posts, and services through an independent recruitment body, Agricultural Scientists' Recruitment Board (ASRB), which is accountable to the ICAR Society.
- ASRB receives funding from the Government of India. The Senior Officers at the ICAR (headquarters). The research set up of the ICAR includes 112 Institutes: 69 Research Institutes, 6 National Bureaux, 23 Project Directorates and Agricultural Technology Application Research Institutes, 14 National Research Centres, and 81 All India Coordinated Research Projects+Network Research Projects.

New Patent Applications Granted

- **Agricultural implements:** Coconut testa removing machine (IN278013); urea molasses mineral block machine (IN283378); coconut chips slicer (IN285418); and Pusa chickpea thresher (IN282047).

- **Animal health/ vaccines/ diagnostics:** Bio-enrichment of cattle feed for better digestibility methods for preparation of antigen from field strain of Mycobacterium aviumsub sp. paratuberculosis and determination of antibody multiple antigenic peptide assay for detection of Peste des Petitis Ruminants (PPR) virus specific antibodies and recombinant antigen based rapid sero-diagnosis of infectious bursal disease (IBD).
- **Cotton and jute textile processes:** Method for the production of cellulose powder from crop residues.
- **Crop production processes:** Improved process for detoxification of Jatropha curcas cake.
- **Dairy Products and processes:** Process for the preparation of ready to constitute and reconstitute makhana, kheermix and chitosan-based bio-preservative mix for buffalo meat.
- **Fish/marine products and processes:** Competitive ELISA for diagnosis and sero-monitoring of microbial infections in cultured freshwater fishes process for differential sero-diagnosis of gram-negative bacterial infections and detection of their toxins in human food, in reference to fish and fish products and method and an apparatus for developing gonadal maturity in carp.
- **Food products and processes:** Device for stripping curry leaf preparation and method of processing of aloe candy from aloe species Jaisalmeri preserve and candy from fruit of toosh (Citrullus colocynthis) process technology for preparation of blended guava leather and method of preparing aloin.
- **Plant-protection methods and processes:** Samfungin: a novel fungicide and the process for making process for the preparation of 5 substituted 1-3,4 oxiadiazole–2 thiols improved neem larvicidal compositions slow or controlled release mosquito larvicidal composition and a process for preparation bio-pesticidal formulation with improved shelf-life super absorbents and the methods of obtaining the same improvement in/or relating to synthesis of o-alkyl derivatives of oxime ethers of piperonal as potential fungicides non-sticky insect trap and insect rearing and collection apparatus.

❑❑❑

Chapter 6

Latest Varieties of Crops

Latest Varieties of Crops

Sr.No.	CROP	VARIETIES	Salient features
1	Mango	Arka Neelanchal kesari	Extra early
2	Custard apple	Arka Neelanchal vikram	Clonal selection
3	Chilli	Arka Neelanchal prabha	
4	Brinjal	Arka Neelanchal shyama	
5	Cross of spine gourd (Momordica subangulata) × Teasel gourd (M.dioica)	Arka Neelanchal shanti	
6	China Aster	Arka Adya (pink),Arka Archana(white)	
7	French bean	Arka sharath	
8	Amaranthas	Arka samraksha, Arka varna	
9	Papaya	Arka prabhat	
10	Crossandra	Arka ambara, Arka kanaka, Arka shravya,Arka shreya	Wilt resistant
11	Mango	Arka udaya (H-12)	Amrapalli × Arka anmol, good keeping quality, TSS-22-25 °Brix
12	Guava	Arka rashmi	Kamsari × Purple local, HDP,red fleshed
13	Gladiolus	Arka ayush	Gold medal × Poonam
14	Gladiolus	Arka manorama	Meera × picardy
15	Marigold	Arka alankar,Arka agni	
16	Gladiolus	Arka Naveen	Purple group
17	Tuberose	Arka nirantara	
18	Lime	Pusa udit, Pusa Abhinav	
19	Rose	Pusa mahak	
20	French marigold	Pusa deep early	
21	Wheat	Pusa akiran	

New varieties of field and horticultural crops

- Early maturing summer Mungbean: **Virat**
- High zinc rice variety: **DRR Dhan 45**
- High protein rice variety: **CR Dhan 310**
- Bacterial leaf blight disease resistant rice varieties: **Pusa 1592 and Punjab Basmati-3**
- Blast disease resistant rice variety: **Pusa 1609**
- Bacterial wilt resistant tomato hybrid: **Arka Samrat**

New Varieties Released by ICAR

- Wheat-Pusa Kiran (HS 542-semi-dwarf variety), Nilgiri Khapli (HW 1098), HDSCW 18 (first variety of the country bred specifically for Conservation Agricultue), HD 3117, Pusa Malvi (HD 4728), HS 562.
- Pearl Millet-Pusa Composite 701
- Mustard –Pusa Double Zero Mustard 31-First canola type, erucic acid less than 2% and glucosinolates < 30 pp
- Pigeon pea -PUSA Arhar 16 (determinate, extra early maturity)
- Sugarcane –C0-0238 (Wonder variety of sugarcane)
- Quality Protein Maize (QPM) -PUSA Breakfast Cereal
- World's first cytoplasmic male sterility-based pigeon pea hybrid-GTH 1
- Low neuro toxin varieties of grass pea-rattan, nirmal, prateek
- Marigold-Pusa Bahar
- High zinc rice variety-DRR Dhan 45
- High protein rice variety-CR Dhan 310
- First ever short duration mung (< 60 days)-Virat (IPM 205-7)
- BLB of rice disease resistant variety-PUSA 1592, Punjab Basmati-3
- Blast disease resistant rice variety-PUSA 109
- **Pusa Betakesari variety of cauliflower-** 1st indigenous bred bio-fortified for beta carotene developed by IARI
- New variety of **pigeon pea** developed by IARI is **Pusa arhar-16** (120 days maturity and determinate in growth)

- World earliest maturity variety of cotton genotype was identified by CICR scientist (2017)- variety name: **Yugank**

AGRI-EXPORT ZONE

Agri-export zone

Sr. No.	State	AEZ Project	Crop	Districts /Area
1.	**West Bengal (6)**	1	Pineapple	Darjeeling, Jalpaiguri, Uttar Dinajpur, Cooch Behar, Howrah
		2	Litchi	Murshidabad, Malda, 24 Pargana(N) and 24 Pargana(S)
		3	Potatoes	Hooghly, Burdhwan, Midnapore (W), Uday Narayanpur, Howrah
		4	Mango	Maldah and Murshidabad
		5	Vegetables	Nadia, Murshidabad and North 24 Parganas
		6	Darjeeling Tea	Darjeeling
2	**Karnataka (4)**	7	Gherkins	Tumkur, Bangalore Urban, Bangalore Rural, Hassan, Kolar, Chitradurga, Dharwad and Bagalkot
		8	Rose Onion	Bangalore (Urban), Bangalore (Rural), Kolar
		9	Flowers	Bangalore (Urban), Bangalore (Rural), Kolar, Tumkur, Kodagu and Belgaum
		10	Vanilla	Districts of Dakshin Kannada, Uttara Kannada, Udupi, Shimoga, Kodagu, Chickamagalur
3	**Uttarakhand (4)**	11	Litchi	Udhamsingh Nagar, Dehradun and Nainital
		12	Flowers	Dehradun, Pantnagar, Udham Singh Nagar, Nainital and Uttarkashi
		13	Basmati Rice	Udham Singh Nagar, Nainital, Dehradun and Haridwar
		14	Medicinal & Aromatic Plants	Uttarkashi, Chamoli, Pithoragarh, Dehradun, Nainital, Haridwar and Udhamsingh Nagar
4	**Punjab (3)**	15	Vegetables (Cabbage Broccoli, Okra, Peas, Carrot, Baby Corn, Green Chillies, Green Beans, Tomato)	Fatehgarh Sahib, Patiala, Sangrur, Ludhiana and Ropar

		16	Potatoes	Singhpura, Zirakpur Distt. Patiala and satellite centres at Rampura Phul, Muktsar, Ludhiana, Jullundur
		17	Basmati Rice	Gurdaspur, Amritsar, Kapurthala, Jalandhar, Hoshiarpur & Nawanshahar
5	**Uttar Pradesh (4)**	18	Potatoes	Agra, Hathras, Farrukhabad, Kannauj, Meerut, Baghpat and Aligarh, Janpad Badaiyun, Rampur, Ghaziabad, and Firozabad
		19	Mangoes and Vegetables	Lucknow, Unnao, Hardoi, Sitapur and Barabanki
		20	Mangoes	Saharanpur, Muzaffarnagar, Bijnore, Meerut, Baghpat and Bulandshahar, Jyotifulenagar
		21	Basmati Rice	Bareilly, Shahjahanpur, Pilibhit, Rampur, Badayun, Bijnore, Moradabad, JB Phulenagar, Sharanpur, Mujjafarnagar, Meerut, Bulandshahar, Ghaziabad and Baghpat
6	**Maharashtra (8)**	22	**Grape & Grapevine**	Nasik, Sangli, Sholapur, Satara, Ahmednagar
		23	**Mangoes**	Rantagiri, Sindhudurg, Raigarh and Thane
		24	**Kesar mango**	Aurangabad, Beed,Jalna, Ahmednagar and Latur
		25	Flowers	Pune, Nasik, Kolhapur and Sangli
		26	Onion	Nasik, Ahmednagar, Pune, Satara, Jalgaon, Sholapur
		27	Pomegranate	Districts of Sholapur, Sangli, Ahmednagar, Pune, Nasik, Osmanabad & Latur
		28	Banana	Jalgaon, Dhule, Nandurbar, Buldhana, Parbhani, Hindoli, Nanded and Wardha
		29	Oranges	Nagpur and Amraoti
7	**Andhra Pradesh (5)**	30	**Mango Pulp & Fresh Vegetables**	Chitoor
		32	Mangoes	Krishna
		33	Gherkins	Districts of Mahboobnagar, Rangareddy, Medak, Karimnagar, Warangal, Ananthapur and Nalgonda
		34	Chilli	Guntur

8	Jammu & Kashmir (2)	35	Apples	Srinagar, Baramula, Anantnag, Kupwara, Kathua and Pulwama
		36	Walnut	Baramulla, Anantnag, Pulwama, Budgam,Kupwara, Srinagar, Doda, Poonch, Udhampur, Rajouri and Kathua
9	Tripura (1)	37	Pineapple	Kumarghat, Manu, Melaghar, Matabari and Kakraban Blocks
10	Madhya Pradesh (5)	38	Potatoes, Onion and Garlic	Malwa, Ujjain, Indore, Dewas, Dhar, Shajapur, Ratlam, Neemuch Mandsaur and Khandwa
		39	Seed Spices	Guna, Mandsaur, Ujjain, Rajgarh, Ratlam, Shajapur and Neemuch
		40	Wheat (including sharbati wheat for Bhopal Zone)	Ujjain Zone (Neemuch, Ratlam, Mandsaur and Ujjain), Indore Zone (Indore, Dhar, Shajapur and Dewas) and Bhopal Zone (Sehore, Vidisha, Raisen, Hoshangabad, Harda, Narsinghpur and Bhopal)
		41	Lentil and Grams	Shivpuri, Guna, Vidisha, Raisen, Narsinghpura, Chhindwara
		42	Oranges	Chhindwara. Jpsjamgabad. Betul
11	Tamil Nadu (4)	43	Cut Flowers	Dharmapuri
		44	Flowers	Nilgiri
		45	Mangoes	Districts of Madurai, Theni, Dindigul, Virudhunagar and Tirunelveli
		46	Cashewnut	Cuddalore, Thanjavur, Pudukottai and Shivaganga
12	Bihar (1)	47	Litchi, Vegetables & Honey	Muzaffarpur, Samastipur, Hajipur, Vaishali, East and West Champaran, Bhagalpur, Begusarai, Khagaria, Sitamarhi, Saran and Gopalganj
13	Gujarat (3)	48	Mangoes & Vegetables	Ahmedabad, Khaida, Anand, Vadodara, Surat, Navsari, Valsad, Bharuch, Narmada
		49	Value Added Onion	Districts of Bhavnagar, Surendranagar, Amreli, Rajkot, Junagadh and Jamnagar Districts
		50	Sesame Seeds	Amerali, Bhavnagar, Surendranagar, Rajkot, Jamnagar
14	Sikkim (2)	51	Flowers (Orchids)& Cherry Pepper	East Sikkim
		52	Ginger	North, East, South & West Sikkim
15	Himachal Pradesh (1)	53	Apples	Shimla, Sirmaur, Kullu, Mandi, Chamba and Kinnaur

16	Orissa (1)	54	Ginger and Turmeric	Kandhamal
17	**Jharkhand (1)**	55	Vegetables	Ranchi, Hazaribagh and Lohardaga
18	**Kerala (2)**	56	Horticulture Products	Thrissur, Ernakulam, Kottayam, Alapuzha, Pathanamthitta, Kollam, Thiruvanthapuram, Idukki and Pallakad
		57	Medicinal Plant	Wayanad, Mallapuram, Palakkad, Thrissur, Ernakulam, Idukki, Kollam, Pathanamittha, Thiruvananthapuram
19	**Assam (1)**	58	Fresh and Processed Ginger	Kamrup, Nalbari, Barpeta, Darrang, Nagaon, Morigaon, Karbi Anglong and North Cachar
20	**Rajasthan (2)**	59	Coriander	Kota, Bundi, Baran, Jhalawar & Chittoor
		60	Cumin	Nagaur, Barmer, Jalore, Pali and Jodhpur

❑❑❑

Chapter 7

Krishi Gyan Mobile Apps

Krishi Gyan Mobile App s - CROPS

1. Rice Xpert

- App developed by National Rice Research Institute, Cuttack in 2017.
- Rice Xpert is developed for the farmers to provide information in real time on insect pests, nutrients, weeds, nematodes and disease-related problems, rice varieties for different ecologies, farm implements for different field and post-harvest operations.
- App as a diagnostic tool in rice fields and make customize queries for quick solution through text, picture and voice.

2. RKMP RiceVocs

- App developed by Indian Institute of Rice Research, Hyderabad, Telangana in 2015.
- Features include 2500 terms related to rice crop arranged in alphabetical manner.
- For extension professionals and other stakeholders, this act as a ready reckoner in getting right meaning about different technical terms.

3. Rice IFC (Insecticide & Fungicide Calculator)

- App developed by ICAR National Centre for Integrated Pest Management (NCIPM), New Delhi in 2017.
- Features include information on chemical and biological insecticides and fungicides.
- Weather based predictions of insect pests and diseases of rice.

4. Rabi Sorghum

- Mobile App developed by Krishi Vigyan Kendra, Gadag, Karnataka in 2015.
- Features include offline and online mode of crop management practices for Rabi sorghum.
- Information about seeds, fertilizers, market prices (only post-harvest technologies) including seed treatment, sowing season and time, pest & diseases with their management.

5. Jwari (Sorghum)

- App developed by Vasantrao Naik Marathwada Krishi Vidyapeeth, Parbhani, Maharashtra in 2017.
- Features include offline and online mode for sorghum cultivation and management.
- Information on sorghum diseases and pests with images.
- App supports Marathi language.

6. Information on Food grains

- App developed by an Agricultural University in association with Ikhedut in 2013.
- Features include offline and online mode information on food grain production, FAQ – for Farmers.
- Uploading of image and video facility.
- App supports Gujarati language.

7. Chanamitra

- App developed by ICAR- Indian Institute of Pulses Research, Kanpur in 2016.
- Features include information related to improved chickpea varieties, crop production and protection technologies for management of insect pest and diseases, post-harvest technologies.
- App supports English and Hindi languages.

8. Cane Adviser

- App developed by ICAR Sugarcane Breeding Institute, Coimbatore, Tamil Nadu in 2017.
- Features include sugarcane varieties, production and protection technologies and fertilizer schedule.
- Query handler helps to raise queries as text or in graphic form that are replied via SMS / email.
- App supports English, Tamil and Hindi languages.

9. Cotton (Kapus)

- Developed by Vasantrao Naik Marathwada Krishi Vidyapeeth, Parbhani, Maharashtra in 2017.
- Features include offline and online mode, information on cotton cultivation for Marathwada (Maharashtra) with useful information on package of practices of cotton, such as pest and disease management, seeds, fertilizer management, irrigation, critical growth stages, etc.
- App supports Marathi language.

10. Soyabean-Gyan

- App developed by ICAR Indian Institute of Soyabean Research, Indore, Madhya Pradesh and TCS (m-Krishi) in 2017.
- Features include seeds, soyabean varieties, their maturity duration, yield potential, suitability in agro-climatic zones, recommended package of practices, controlling measures for managing insect-pests and diseases and weed management.
- App supports Hindi language.

11. E-Thilhan

- App developed by ICAR-Indian Institute of Oilseeds Research, Hyderabad, Telangana in 2017.
- Features include information on sesame and sunflower viz., varieties, suitable cropping systems, pests, diseases, AICRP centres and commodity markets.
- App supports English and Telugu languages.

12. ICAR IIOR

- App developed by ICAR-Indian Institute of Oilseeds Research, Hyderabad, Telangana in 2017.
- Features include information on sunflower, safflower, castor and sesame farming viz. agronomic practices, production technology, crop management, insect management, disease management, weeds management, food uses and health benefits.
- App supports English language.

13. ICAR IIOR Castor

- App developed by ICAR-Indian Institute of Oilseeds Research, Hyderabad, Telangana in 2017.
- Features include information on castor viz., agronomic practices, cultivars, state-wise preferred varieties and hybrids, inter-cropping systems recommended for different states, insect pests, diseases, AICRP centres and commodity markets, important APMC's trading of castor.
- App supports English language.

14. ICAR IIOR Sunflower

- App developed by ICAR-Indian Institute of Oilseeds Research, Hyderabad, Telangana in 2017.
- Features include information on sunflower viz., agronomic practices, cultivars, state-wise preferred varieties and hybrids, inter-cropping systems recommended for different states, insect pests, diseases, AICRP centres and commodity markets, important APMC's trading of sunflower.
- App supports English language.

15. ICAR IIOR Safflower

- App developed by ICAR-Indian Institute of Oilseeds Research, Hyderabad, Telangana in 2017.
- Features include information on safflower viz., agronomic practices, cultivars, state-wise preferred varieties and hybrids, inter-cropping systems recommended for different states, insect

pests, diseases, AICRP centres, commodity markets and APMC's trading safflower.

- App supports English language.

16. ICAR IIOR Sesame

- App developed by ICAR-Indian Institute of Oilseeds Research, Hyderabad, Telangana in 2017.
- Features include information on sesame viz., agronomic practices, cultivars, state-wise preferred varieties and hybrids, inter-cropping systems recommended for different states, insect pests, diseases, AICRP centres, commodity markets and APMC's trading sesame.
- App supports English language.

17. ICAR IIOR Bio-Controls

- App developed by ICAR-Indian Institute of Oilseeds Research, Hyderabad, Telangana in 2017.
- Features include information on Bacillus thuringiensis (Bt) which is a gram-positive ubiquitous soil bacterium producing an insecticidal toxin, advantages, source of availability, varieties and suitable cropping systems.
- Information on Trichoderma seed treatment benefits, disease control and dosage.
- App supports English language.

18. Fodder App

- App developed by Kerala Veterinary and Animal Sciences University, Wayanand, Kerala in 2017.
- Features include information on different varieties of fodder and their cultivation method and productivity with pictures.
- Provides end-to-end information about fodder viz., varieties, soil and weather suitable, fertilizer required, cultivation practices, nutrition content, harvesting and yield, etc.
- App supports English language.

19. PDKV Weed Manager

- App developed by Dr. Panjabrao Deshmukh Krishi Vidyapeeth, Akola, Maharashtra in 2017.
- Features include offline information on weed management of crops viz., cotton, soyabean, oilseeds, paddy, sugarcane, wheat, sorghum.
- Facilities of uploading of images, information and feedback.
- App supports Marathi language.

20. Weed Manager

- App developed by ICAR Directorate of Weed Research, Jabalpur, Madhya Pradesh in 2017.
- Some features are offline and some query-based features are online, allow users to scout crop name and identify common dominated weeds of that particular crop along with their control measures.
- Crop-wise weed control recommendations for Kharif, Rabi and Zaid.
- App supports English language.

21. Nematode Info

- App developed by ICAR Indian Agricultural Research Institute, New Delhi in 2017.
- Features include various nematodes that affect the crops and their management.
- App considers only the nematodes that are found in rice and wheat crops.
- App supports English and Hindi languages.

22. Micromitra

- App developed by ICAR- National Bureau of Agriculturally Important Micro-organisms, Mau, Uttar Pradesh in 2017.
- Features include information on various micro-organisms, bio-fertilizers, bio-control agents and decomposers.

- Information about liquid formulation of NPK providing bacteria for various crops, advantages, Application, dosages, precautions and cost.
- App supports English and Hindi languages.

23. E-Mausam HAU Krishi Mausam Seva

- App developed by CCS Haryana Agricultural University, Hissar in 2017.
- Features include information related to agriculture like district weather forecasts, present weather, weather-based crop advisory and crop package and practices
- App supports Hindi language.

24. VNMKV (Dry land Cultivation)

- App developed by Vasantrao Naik Marathwada Krishi Vidyapeeth, Parbhani, Maharashtra in 2017.
- Features include package of practices of crops for dry land cultivation for sustainable productivity under changing climate situation, importance of soil and water conservation, improved agricultural implements, crop cultivation and weed management in dry land cultivation.
- Information on weather and dry land cultivation in Marathwada.
- App supports Marathi language.

25. Mobile Farm Solutions(Q & A)

- App developed for KVKs by Department of Agriculture, Government of Meghalaya and Directorate of Agriculture in 2017.
- Feature include solution of problems related to crops by the Scientist/Subject Matter Specialist.
- App helps the farmers to report problems on the crops, soil, pests, etc to the Scientist or SMS of KVKs in their respective districts.
- App supports English language.

26. Krishi Sparsham

- App developed by KVK Pathanamthitta District, Kerala in 2016.
- Features include online and offline information on pest and diseases management of major crops, such as Paddy, Banana, Vegetables and Coconut in Kerala.
- App supports Malayalam language.

27. Kheti Gyan

- App developed by Krishi Vigyan Kendra, Jalandhar, Punjab in 2017.
- Features include photographs of crop information and farm machinery, major crop information in offline mode.
- App supports English and Punjabi languages.

28. Agrowbook

- App developed by ICT based Company in 2015 in association with ICAR - Academy of National Agricultural Research Management, Rajendranagar, Hyderabad, Telangana.
- Features include seeds, fertilizers, market prices weather forecasting and equipment/machineries, nursery plants, organic/herbal farm inputs.
- Photos of the crop can be uploaded which is answered by the expert.
- App supports English language.

29. mAgIDS

- App developed by Punjab Agricultural University, Ludhiana, Punjab in 2014.
- Features include offline and online mode, crop disease database, send images of diseased crops to experts, quick replies for queries and any other information.
- App supports Hindi, English and Punjabi languages.

30. Agri eXpert System

- App developed by University of Agricultural Sciences, Department of Plant Pathology, College of Agriculture, Bengaluru, Karnataka in 2014.
- Features include online crop disease database, send images of diseased crops to experts, quick replies for queries and any other information embedded.
- App supports English and Kannada languages.

31. Khetiyok

- App developed by Krishi Vigyan Kendra, Dhemaji, Assam in 2017.
- Features include crop disease database, send images of diseased crops to experts, quick replies for queries and any other information.
- App supports Assamese language.

32. Malwa Fasal

- App developed by Krishi Vigyan Kendra, Ujjain, Madhya Pradesh in 2017.
- Features include information about seeds and fertilizers and package of practices in audio form through Google play console.
- App supports English and Hindi languages.

33. Kheti Sewa

- App developed by Krishi Vigyan Kendra, Amritsar, Punjab in 2017.
- Features include information about KVK present and past events, package of practices, database, faculty, etc. interact with the farmers and to obtain feedback and suggestions.
- App supports English and Punjabi languages.

34. Farm Calculators

- App developed by University of Agricultural Sciences, Bengaluru, Karnataka in 2015.
- Features include offline use of exact quantity of seeds and fertilizers. The App is useful in saving inputs and provides accurate quantity to be Applied in crops.
- App supports English, Hindi, Kannada, Tamil, Telugu, Marathi, Bengali and Gujarati languages.

35. DEE PDKV Transfer of Technology

- Developed by Directorate of Extension Education, Dr. PDKV, Akola, Maharashtra in 2017.
- Features include offline and online mode of information on crops (cotton, soyabean, oilseeds, paddy, sugarcane, wheat, sorghum and agricultural implements).
- Feedback or contact for query to ATIC on toll free helpline.
- App supports English and Marathi languages.

36. Phule Krishidarshani

- App developed by Mahatma Phule Krishi Vidyapeeth, Rahuri, Maharashtra in 2017.
- Provide information about crops, instrument and modern technology for additional growth of agricultural production.
- Ability to upload images, availability of public forum groups for discussions and sharing of new ideas and practices of crops, etc.
- Queries are answered by E-mail, SMS and Phone call.
- App supports Marathi language.

37. Kisan Sahayak Ropar

- App developed by Kris1hi Vigyan Kendra, Ropar, Punjab in 2017.
- Features include offline crop disease database, send images of diseased crops to experts, quick replies for queries and any other information, IMD weather for Ropar embedded, seeds portal available.
- App supports English and Punjabi languages.

38. Kisan Sahayak Fatehgarh Sahib

- App developed by Krishi Vigyan Kendra, Fatehgarh Sahib, Punjab in 2017.
- Features include list of available seeds at KVK, current market value of crops and weather forecasting, send images of diseased crops to experts, quick replies for queries and any other information, IMD weather for Fatehgarh Sahib embedded, seeds portal.
- App supports English and Punjabi languages.

39. AgroTech VNMKV

- App developed by Vasantrao Naik Marathwada Krishi Vidyapeeth, Parbhani, Maharashtra in 2017.
- Features include both online and offline information to provide complete details about farming and its technologies, IMD weather for Parbhani embedded, seeds portal available.
- App supports Marathi language.

40. IPM VNMKV Parbhani

- App developed by Vasantrao Naik Marathwada Krishi Vidyapeeth, Parbhani, Maharashtra in 2017.
- Features include offline and online mode of crop disease database, send images of diseased/pests crops to experts, replies for queries and any other information.
- App supports Marathi language.

41. Malda Krishi Vigyan Kendra

- App developed by Malda Krishi Vigyan Kendra, Ratua, Malda, West Bengal in 2017.
- Features include information on seeds, live market price and weather forecast. Provide information on vocational training to the practising farmers including farm women, youth and extension functionaries in improved technologies in the field of agriculture, animal husbandry, fisheries and other allied enterprises.
- App supports English, Bengali, Hindi and Nepali languages.

42. KVK Faridkot

- App developed by KVK Faridkot, Punjab Agricultural University, Ludhiana, Punjab in the year 2018.
- Features include information and activities of KVK, Faridkot, package of practices of PAU, online report of soil and water testing of PAU, feedback on training course, aridkot action plan and trainings of KVK, profile of Faridkot and farmer query.
- App supports English language.

Krishi Gyan Mobile App s – HORTICULTURE

1. ICAR-NRCP Pomegranate

- App developed by National Research Centre on Pomegranate, Sholapur, Maharashtra in 2017.
- Features include scientific pomegranate production practices and educate pomegranate growers.
- Supports English, Hindi and Marathi languages.

2. Sholapur Anar

- App developed by ICAR-National Research Centre on Pomegranate, Sholapur, Maharashtra in 2017.
- Features include information on pomegranate viz., nursery and orchard establishment, production of quality planting material, nutrient management, water management, disease and pest management, post-harvest management in pomegranate, process and value addition, current market rates availability and about NRCP, facilities & laboratories information including photo gallery and daily weather data.
- App supports Hindi, Marathi and English languages.

3. ICAR-NRCL

- App developed by ICAR National Research Centre on Litchi, Muzaffarpur, Uttar Pradesh in 2017.
- Features include crop disease database, send images of diseased crops to experts, replies for queries and any other information,

IMD weather Muzaffarpur embedded, ongoing research projects, and the achievements in litchi R&D.

- App supports English and Hindi languages.

4. Grapes DSS

- App developed by ICAR National Research Centre for Grapes, Pune, Maharashtra in 2017.
- App educates grape growers about scientific grape production practices
- App supports English language.

5. Mango Cultivation IIHR

- App developed by ICAR-Indian Institute of Horticultural Research, Bengaluru, Karnataka in 2016.
- Features include offline Mango crop production viz., soil & climate requirement, propagation, spacing, planting, training & pruning, INM, irrigation and harvesting.
- The crop management aspects comprise of disease management affecting mango crops, viz., anthracnose, blossom blight, leaf blight, powdery mildew, dieback, etc., and the pest management modules comprises of infestation of fruit fly, mango hopper, stone weevil, mealy bug, shoot borer, stem borer, etc.
- App supports English language.

6. *Aam ki Suraksha* - ICAR Patna

- App developed by ICAR- Research Complex for Eastern Region, Patna, Bihar in 2018.
- Features include offline information on pests and disease of mango.
- App supports Hindi language.

7. Raw Mango Products

- App developed by ICAR- Central Institute for Subtropical Horticulture, Lucknow, Uttar Pradesh in 2017.

- Features include making and relishing Indian styled raw mango products. Information on reducing post-harvest losses of raw mangoes, value added product of Mango.
- App supports Hindi language (Audio feature also available).

8. Ripe Mango Products

- App developed by ICAR Central Institute for Subtropical Horticulture, Lucknow, Uttar Pradesh in 2017.
- Features include preparation of ripe mango product, such as juices, nectars, drinks, jams, fruit cheese, ice cream, yoghurts, puddings, filling, bakery products and bars.
- App supports Hindi language (Audio feature also available).

9. Citrus Cultivation

- App developed by Vasantrao Naik Marathwada Krishi Vidyapeeth, Parbhani, Maharashtra in 2014.
- Features include offline complete information about citrus cultivation, processing and its technologies.
- App supports Marathi language.

10. CCRI-Citrus

- App developed by ICAR Central Citrus Research Institute, Nagpur, Maharashtra in 2017.
- Features include effective and citrus cultivation practices, monthly calendar of operations, etc. Provides online and offline information to the farmers and other clientele groups.
- App supports English, Hindi and Marathi languages.

11. FC in Asom

- App developed by Krishi Vigyan Kendra, Dibrugarh, Assam Agricultural University in 2017.
- Features include package of practices of fruit crops (pineapple, lemon, papaya, mango, guava, arecanut, cashewnut, coconut, sapota, litchi, betelvine and jackfruit).

- Farmers can check market price of different crops, as it is linked with Agri-market mobile App.
- App supports English language.

12. Fruit Crops

- App developed by ICAR-Indian Institute of Horticultural Research, Bengaluru, Karnataka in 2017.
- Features include improved varieties of fruits (mango, papaya, grapes, guava, pomegranate, custard App le, fig, jackfruit, pummelo & underutilized fruits) developed by the institute.
- App supports English language.

13. Horticultural Package in Assam

- App developed by Directorate of Extension Education, Assam Agricultural University, Jorhat in 2017.
- Features include package of practices of horticulture crops.
- App supports Assamese language.

14. Tomato Cultivation IIHR

- App developed by ICAR-Indian Institute of Horticultural Research, Bengaluru, Karnataka in 2017.
- Features include offline tomato crop production viz., land preparation, use of bio-agents, nursery raising and seed rate, transplanting, drip irrigation, fertilizers, inter-cultivation, IPM, disease and pest management.
- App supports English language.

15. Sabjee Gyan

- App developed by ICAR-Indian Institute of Vegetable Research, Varanasi, Uttar Pradesh in 2017.
- Features include vegetable production related information, best cultivation practices on important vegetable crops, insect and disease management information and weather services.
- App supports English and Hindi languages.

16. Tuber Guru

- App developed by ICAR Central Tuber Crops Research Institute, Thiruvananthapura, Kerala in 2018.
- Features include information on seeds, fertilizers, varieties, agro techniques, manures and fertilizers, intercultural operations, intercrops, pests and diseases of important tropical tuber crops like cassava, sweet potato, yams, elephant foot yam and taro.
- App supports English and Malayalam languages.

17. Oyster Mushroom

- App developed by Krishi Vigyan Kendra, Kamrup in 2018.
- Features include scientific cultivation practices and production technology of Oyster mushroom in Assam.
- App supports Assamese language.

18. ICAR-Mushroom

- App developed by ICAR Directorate of Mushroom Research, Solan, Himachal Pradesh in 2016.
- Features include nutritional and medicinal facts, photo of cultivated edible and poisonous mushrooms with description, technologies developed by ICAR-DMR, Mushroom products photographs and its ingredients, cultivation techniques of different types of mushroom.
- In-App Message service provides crop advisory and news-events updates.
- App supports English language.

19. Oil Palm Cultivation Practices

- App developed by ICAR Indian Institute of Oil Palm Research, Pedavegi, Andhra Pradesh in 2016.
- Features include offline recommended practices of oil palm cultivation in India. Management of oil palm plantations.
- App supports English, Hindi and Telugu languages.

20. Oil Palm Nutrient Management

- App developed by ICAR Indian Institute of Oil Palm Research, Pedavegi, Andhra Pradesh in 2016.
- Features include offline information on symptoms of nutrient deficiencies viz. nitrogen, potassium, phosphorus, boron, magnesium, iron, copper, zinc, manganese and other disorders in oil palm along with management practices.
- App supports English, Hindi and Telugu languages.

21. Oil Palm Pest Management

- App developed by ICAR Indian Institute of Oil Palm Research, Pedavegi, Andhra Pradesh in 2016.
- Features include offline information on symptoms of pest infestation viz., rhinoceros beetle, scales, mealy bugs, slug caterpillar, bag worm, chafer beetles, termites, leaf webworm, birds, rats and wild animals in oil palm and their management.
- App supports English, Hindi and Telugu languages.

22. Oil Palm Disease Management

- App developed by ICAR Indian Institute of Oil Palm Research, Pedavegi, Andhra Pradesh in 2016.
- Features include offline information on symptoms of pest infestation viz., rhinoceros beetle, scales, mealy bugs, slug caterpillar, bag worm, chafer beetles, termites, leaf webworm, birds, rats and wild animals in oil palm and their management.
- App supports English, Hindi and Telugu languages.

23. E-kalpa

- App developed by ICAR- Central Plantation Crops Research Institute, Kasaragod, Kerala in 2016.
- Features include offline cloud based interactive mobile Application for plantation crop (coconut, arecanut and cocoa) farmers and stakeholders.
- App supports English, Malayalam, Kannada, Hindi languages (at present) and Bengali, Tamil, Telugu, Assamese (in progress).

24. Turmeric Cultivation

- App developed by Vasantrao Naik Marathwada Krishi Vidyapeeth, Parbhani, Maharashtra in 2016.
- Features include offline information about turmeric cultivation and processing technologies.
- App supports Marathi language.

25. Seed Spices Info

- App developed by ICAR National Research Centre on Seed Spices, Ajmer, Rajasthan in 2017.
- Features include crop production, crop improvement, crop protection, value addition and transfer of technology.
- Crop disease database, send images of diseased crops to experts, quick replies for queries and any other information.
- App supports Hindi and English languages.

26. ICAR IISR Black pepper

- App developed by ICAR Indian Institute of Spices Research, Kozhikode, Kerala in 2017.
- Features include information on cultivation, management, plant protection, planting material availability and fertilizer recommendation of Black Pepper.
- App supports English, Hindi and Telugu languages.

27. ICAR-DMAPR

- App developed by ICAR Directorate of Medicinal and Aromatic Plants Research, Anand, Gujarat in 2016.
- Features include information of medicinal and aromatic crops identification, cultivation techniques and good agriculture practices.
- App supports English language.

Krishi Gyan Mobile App s - VETERINARY

1. IVRI-Shukar Palan App (Pig Farming)

- App developed by ICAR- Indian Veterinary Research Institute, Bareilly, Uttar Pradesh in 2018.
- Features include offline to impart scientific knowledge and skills to the graduating veterinarians, field veterinary officers, developmental organizations and entrepreneurs for promoting commercial pig farming.
- Specific e-mail support for users.
- App supports I lindi language.

2. IVRI-Pashu Prajanan (Animal Reproduction)

- App developed by ICAR-Indian Veterinary Research Institute, Bareilly, Uttar Pradesh in 2017.
- Features include offline reproductive diseases/disorders in cattle and buffaloes viz., anoestrus, repeat breeding, silent estrus/silent heat, uterine torsion, dystocia, abortion, uterine prolapse, retention of foetal membranes/ placenta, metritis, brucellosis, campylobacteriosis and IBRIPV which leads to economic losses.
- App supports English, Hindi, Punjabi, Bangla, Assamese, Gujarati and Tamil languages.

3. E-Pashupalan

- Mobile App developed by Lala Lajpat Rai University of Veterinary and Animal Sciences, Hisar, Haryana in 2014.
- Features include offline information about dairy farming, sheep & goat farming, Pig farming, cow and buffalo breeds, housing management, nutrition management, reproductive management, artificial insemination, signs of heat, etc., disease information, prevention and control of different diseases, main clinical signs and symptoms of different diseases of animals, etc., vaccination and clean milk production.
- App supports Hindi language.

4. Info Equine

- App developed by ICAR-National Research Centre on Equines, Hisar, Haryana in 2017.
- Features include offline knowledge in various aspects of equines with regard to breeds, management, nutrition, diseases, artificial insemination, pregnancy diagnosis, etc.
- App supports English and Hindi languages.

5. LDF-Livestock Disease Forewarning

- App developed by ICAR National Institute of Veterinary Epidemiology and Disease Informatics (NIVEDI), Bengaluru, Karnataka in 2017.
- Features include offline details of clinical samples to be collected in case of outbreaks of the listed diseases for laboratory confirmation.
- App supports English language.

6. Hoof Care

- App developed by Kerala Veterinary and Animal Sciences University, Wayanand, Kerala in 2017.
- Features include farmers to determine and maintain the hoof to be trimmed for healthy production.
- App supports English language.

7. Cattle Dentition

- App developed by Kerala Veterinary and Animal Sciences University, Wayanand, Kerala in 2017.
- Features include information to assess the age of cow, goat and dogs to support the farmers while purchasing the animals.
- App supports English language.

8. TANUVAS Vet Micro

- Mobile App developed by Department of Veterinary Microbiology, Madras Veterinary College, Tamil Nadu in 2017.

- Features include practical Veterinary Microbiology course as per MSVE, 2016 syllabus.
- App supports English language.

9. Feed Calculator

- App developed by Tamil Nadu Veterinary and Animal Sciences University in 2017.
- Features include vernacular language which calculates the weight of milch animal in kilograms based on length and girth of the animal, fat percentage, milk yield and calculates the feed requirements.
- App supports English and Tamil languages.

10. Training Calendar

- App developed by Tamil Nadu Veterinary and Animal Sciences University, Chennai in 2017.
- Features include capacity building programmes offered at 20 Veterinary University Training and Research Centres, 3 Farmers Training Centres and 3 Krishi Vigyan Kendras with user friendly menus to browse for centre-wise, district-wise or subject-wise programmes, registration and set up of reminder call.
- App supports English and Tamil languages.

Krishi Gyan Mobile App s – DAIRY

1. Indigenous Dairy Products

- App developed by Kerala Veterinary and Animal Sciences University, Wayanand, Kerala in 2017.
- Features include different milk products, their commercial production, manufacturing technology and inputs required.
- App supports English language.

2. Eco Dairy

- App developed by ICAR-National Dairy Research Institute, Karnal, Haryana in 2017.

- Features include offline information regarding the existing environment friendly dairy farming practices i.e. breeding, feeding, health care and management. Quick replies for queries and any other information.
- App supports Hindi and English languages.

3. Food Safety

- App developed by Kerala Veterinary and Animal Sciences University, Wayanand, Kerala in 2017.
- Features include the issues regarding safe food production and important practices to be followed ensuring safe food.
- App supports English language.

4. ICAR-NIANP Feed Chart

- App developed by ICAR National Institute of Animal Nutrition and Physiology, Bengaluru, Karnataka in 2018.
- Features include formulating ration for crossbred dairy cow and milch buffalo.
- App supports English, Hindi, Kannada, Tamil and Telugu languages.

5. Dairy Kannada

- App developed by Karnataka Veterinary, Animal and Fisheries Sciences University, Bidar & Jayalakshmi Agrotech Limited in 2016.
- Features include important cattle and buffalo breeds, breeding management, selection of dairy animal, housing management, feeding management, general care and management of calves, milch animal varieties, diseases, clean milk production, economic dairying, vaccination, feed calculator and record keeping.
- App supports Kannada and Telugu languages.

6. Fodder Kannada

- App developed by Department of Veterinary & A.H. Extension, Veterinary College, Bengaluru, Karnataka Veterinary, Animal and Fisheries Sciences University, Bidar & Jayalakshmi Agrotech Limited in 2016.

- Provides a repository of information both in audio & visual form, regarding 16 fodders suitable for Karnataka. The soil content, agronomical practices, weather conditions suitable for each fodder, nutrient requirement, nutrient requirement calculator, and expenditure regarding each fodder are provided.
- App supports Kannada language.

Krishi Gyan Mobile App s – POULTRY

1. Farmer Friendly Poultry

- App developed by Sher-e-Kashmir University of Agricultural Sciences and Technology, Jammu in 2018.
- Features include offline content to calculate average weight of the flock, total feed consumed, total weight gained, feed conversion ratio, mortality rate, total input cost (feed, medicines, miscellaneous), total revenue generated and total profit/loss from the flock.
- Feature, such as ability to upload images, availability of public forum groups for discussions and sharing of new ideas and practices, etc. Queries are answered by E- mail, SMS and Phone call.
- App supports English language.

Krishi Gyan Mobile App s – FISHERIES

1. Vanami Shrimp App

- App developed by ICAR Central Institute of Brackish Water Aquaculture, Chennai, Tamil Nadu in 2017.
- Features include management practices of pacific white shrimp farming, such as pond preparation, seed selection and stocking, feeding and feed management, soil and water quality management, health management, regulation, food safety and record keeping.
- App supports Hindi, Telugu and Tamil languages.

2. mKRISHI Fisheries

- App developed in collaboration with Tata Consultancy Services (TCS) Innovation Lab – Mumbai, ICAR- Central Marine Fisheries Research Institute and Indian National Centre for Ocean Information Services (INCOIS), Hyderabad in 2017.
- Features include remote sensing data received from NOAA satellites, sea surface temperature and presence of phytoplankton which form the food of several fish species.
- Fishermen can use this service to plan their fishing trip and venture into sea.
- Saving upto 30% diesel cost.
- mKRISHI®Fisheries has been selected as one of the Top20Social Innovation in a national contest developed by Ministry of External Affairs, Govt. of India and NITI Aayog during Pravasi Bharatiya Divas, 2017.
- App supports Marathi and English languages.

3. ICAR-CIFT

- App developed by ICAR Central Institute of Fisheries Technology, Kochi, Kerala in 2017.
- Features include information on basic and strategic research in fishing and processing.
- App supports English language.

Krishi Gyan Mobile Apps - NATURAL RESOURCE MANAGEMENT

1. Plant Nutrition

- App developed by Vasantrao Naik Marathwada Krishi Vidyapeeth, Parbhani, Maharashtra in 2017.
- Features include plant nutrition and elements present in the soil fertilizers used for the plants.
- App supports Marathi language.

2. Soil Nutrient Manager

- App developed by ICAR Research Complex for Eastern Region, Patna, Bihar in 2018.
- Features include fertilizer recommendations based on native soil fertility status and nutritional requirements of the crop to be grown.
- App supports English and Hindi languages.

3. Nutrient Deficiency Diagnoser and Manager for Apple (NDDMA)

- App developed by ICAR Central Institute of Temperate Horticulture, Srinagar, Jammu & Kashmir.
- Features include management for curing nutrient deficiency in Apple, to optimize yield and reduce yield losses.
- App supports English, Hindi and Urdu languages.

4. MGR-Portal (Microbial Genetic Resource)

- App developed by ICAR National Bureau of Agricultural Important Microorganisms, Mau, Uttar Pradesh in 2017.
- Features include search options for general search of microbial accessions by accession number or by name in NAIMCC database, list of microbial accessions in core collection developed based on agriculturally important traits, procurement of microbial cultures from NAIMCC and opinion for query related to microbial germplasm at ICAR-NBAIM, Mau, Uttar Pradesh.
- App supports English language.

5. Urvara

- App developed by ICAR Research Complex for Eastern Region Patna, Bihar in 2018.
- Features include fertilizer recommendation for a crop based on soil test report/soil health card.
- App supports English language.

6. Fertilizer Calculator - Goa

- App developed by ICAR – Central Coastal Agricultural Research Institute, Goa in 2015.
- Features include offline soil test-based fertilizer recommender (STFR) of crops.
- Get calculations according to the area of farm or the number of plants/trees.
- Specially customized for Goan farms. (For places outside Goa, try out other Fertilizer Calculator App).
- App supports English language.

7. Digital Soil Health Card

- Mobile App developed by Krishi Vigyan Kendra, Sambhalpur, Odisha in 2016.
- Features include soil health card of the farmer with recommendation for use of fertilizer for different crops.
- Image based agro advisory to the farmers.
- App supports English language.

8. Saur Shakti ICAR

- Mobile App developed by ICAR Research Complex for Eastern Region, Patna, Bihar in 2018.
- Facilitates selection of appropriate pump size considering the user input on crops, plot areas, water table, and drawdown, etc.
- Features also include number of plots to be irrigated in one day and discharge requirement.
- App supports English language.

9. PhuleJal

- App developed by Mahatma Phule Krishi Vidyapeeth, Rahuri, Maharashtra in 2017.
- Offline estimates the evapotranspiration (ETr) by different standardized methods for the specific location by fetching the required input weather data from the IMD.
- App supports English and Marathi languages.

10. PIS (Phule Irrigation Scheduler)

- App developed by Mahatma Phule Krishi Vidyapeeth, Rahuri, Maharashtra in 2017.
- App provides decision-making support for irrigation scheduling for crops based on its growth stage, soil type, the local weather condition, type of irrigation method and efficiency of the irrigation system and time of operation of pump/system for surface, sprinkler and drip irrigation methods for different crops.
- App supports English and Marathi languages.

11. VNMKV Ground Water Enhancement

- App developed by Vasantrao Naik Marathwada Krishi Vidyapeeth, Parbhani, Maharashtra in 2017.
- Features include ground water enhancement through artificial recharging and farm pond for rainwater harvesting.
- App supports Marathi language.

12. mKRISHI Paws - IISWC

- App developed by ICAR-Indian Institute of Soil and Water Conservation, Dehradun, Uttarakhand in collaboration with Tata Consultancy Services Limited in 2016.
- Features include online mode Personalized Advisory on Water and Soil (PAWS) – an innovative mobile based extension service to the farmers in remote and hilly region.
- App supports English and Hindi languages.

13. GypCal - Sodic Soil Reclamation

- App developed by ICAR Central Soil Salinity Research Institute, Karnal, Haryana in 2017.
- Features include chemical reclamation of sodic soil for optimizing crop production in Indo-Gangetic plains by calculating the gypsum requirement in bags of 50 Kg. Calculates total depth of water required for leaching to flush out salts.
- App supports English and Hindi languages.

14. Havaamaana-Krishi

- App developed by University of Agricultural Sciences, Dharwad, Karnataka in 2016.
- Features include offline information on weather, short range weather forecast and agromet advisory for seven districts under the jurisdiction of UAS Dharwad in north Karnataka to save crops from adverse weather conditions as well as animal husbandry.
- Linked to the official website of AICRP on Agro-meteorology, Vijayapura Centre, *www.agrometbijapur.org* which includes daily weather data, short range weather forecast, and medium range weather forecast based Agromet Advisory.
- App supports English and Kannada languages.

15. MAA Jharkhand Weather

- App developed by Birsa Agricultural University, Ranchi, Jharkhand in 2017.
- Features include offline content, disseminate weather information on 14 parameters viz. rainfall, maximum temperature, minimum temperature, dew point temperature, soil temperature, soil moisture, surface runoff, underground runoff, downward short-wave flux at ground surface, total cloud cover, maximum relative humidity, minimum relative humidity, wind speed, wind direction on daily (3 days duration) and hourly (for 6 hours) basis.
- App supports English, Hindi and Nagpuri languages.

16. LRISGoa

- App developed by ICAR National Bureau of Soil Survey and Land Utilization Planning,

17. LRIS GOA in 2018

- Features include visualizing, disseminating, sharing and data mining of the land resources in a digital manner.
- App can locate the user's location on hierarchical drop-down selection basis or else the GPS enabled location tracking.
- App supports English language.

18. PCZ Mapper

- App developed by ICAR National Bureau of Soil Survey and Land Utilisation Planning, LRIS GOA in 2018.
- Features include easy visualization, dissemination and sharing of database on crop suitability and potential crop zone of the country.
- App supports English and Hindi languages.

Krishi Gyan Mobile Apps – INTEGRATED

1. Krishi Vigyan

- App developed by Krishi Vigyan Kendra, Amadalavalasa, Srikakulam, Andhra Pradesh in 2016.
- Features include modern scientific management practices for Agriculture & Horticulture crops grown in Andhra Pradesh along with photographs.
- App supports Telugu language.

2. Kisan Mitra

- App developed by Krishi Vigyan Kendra, Navsari, Gujarat in 2014.
- Features include package of practices for agricultural and horticultural crops of south Gujarat, package of practices for horticultural crops, livestock information and important technologies of Navsari Agricultural University and FAQ.
- App supports Gujarati language.

3. Smart NE Kisan

- App developed by Krishi Vigyan Kendra, Anjaw, ICAR-RC for NEH Region, Arunachal Pradesh Centre, Basar in 2016.
- Features include solutions relevant to fishery, livestock, agriculture and provide quick information to them even offline.
- App supports English language.

4. Agriculture: FEM@Mobile

- App developed by Krishi Vigyan Kendra, Malappuram, Kerala in 2015.
- Features include online and offline, crop cultivation, plant protection, organic inputs, agro chemicals, expert support and contact directory.
- App supports English language.

5. Annada

- App developed by Orissa University of Agriculture and Technology, Odisha in 2017.
- Features include online management of livestock, poultry, etc. along with agro-advisory released by OUAT.
- Audio advisory has also been included for the illiterate farmer.
- App supports English and Oriya languages.

6. KVK App

- App developed by Indian Agricultural Statistics Research Institute, New Delhi in 2016.
- App provides information about all the KVKs by selecting state and district, all facilities provided by KVK.
- Agro-meteorological advisory in English/local language.
- Access agro-commodity prices from eNAM portal.
- App supports English language.

7. KVK Barpeta

- App developed by KVK Barpeta (in collaboration with KVK, Kokrajhar) in 2017.
- Features include offline package of practices of different crops, scientific rearing of livestock, plant protection and cropping sequence.
- App supports Assamese language.

8. KVK Mobile App

- App developed by Krishi Vigyan Kendra, Badwani in 2016.
- Features include online information about seeds, fertilizers and complete package and practices of major agricultural and horticultural crops.
- App supports English and Hindi languages.

9. KVK online AgriMart

- App developed by Darjeeling, West Bengal KVK & farming community in 2017.
- Features include online market platform to the farming community of Darjeeling district free of cost.
- App supports English and Nepali languages.

10. CGKV App

- App developed by Chhattisgarh Kamdhenu Vishwavidyalaya, Durg in 2017.
- Features include information regarding research, extension, and academic activities of the University.
- App supports English language.

Chapter 8

Animal Husbandry & Dairy Technology

Rashtriya Gokul Mission

- "Rashtriya Gokul Mission" has been launched in Dec., 2014 with an outlay of Rs. 2025 crore for development and conservation of indigenous breeds through selective breeding especially for genetic upgradation of 11.3 crore low milk yield non-descript bovine population. Project period is from 2014-15 to 2019-20.
- The Rashtriya Gokul Mission is very timely in that sense the cow as an income asset will become more productive and useful to the small and marginal farmer who largely own more than 90% of the poor productive cows.
- With increasing mechanisation of Agriculture, the demand of draught purpose breeds of cattle is reducing. These breeds are being preserved at Gokul Grams, National Kamdhenu Breeding Centres and also by creating Nucleus Herd through Embryo Transfer Technology and IVF.

Objectives of this scheme

- To promote indigenous cattle rearing and conservation in a scientific and holistic manner.
- To enhance productivity of indigenous breeds and increase economic returns from animal products in a sustainable manner.
- To propagate high genetic merit bulls of indigenous breeds.
- To enhance breeding coverage through strengthening breeding network.

The Rashtriya Gokul Mission also envisages establishment of integrated cattle development centres 'Gokul Grams to develop indigenous breeds including upto 40% non-descript breeds:

- To promote indigenous cattle rearing and conservation in a scientific manner.
- To propagate high genetic merit bulls of indigenous breeds.
- To optimize modern Farm Management practices and promote Common Resource Management.
- To utilize animal waste in economical way i.e. Cow Dung, Cow Urine.

National Kamdhenu Breeding Centres (NKBC)

For development, conservation and preservation of Indigenous Breeds, two National Kamdhenu Breeding Centres are being be set up, as a Centre of Excellence to develop and conserve Indigenous Breeds in a holistic and scientific manner. A Nucleus Herd of all the Indigenous Bovine Breeds (41 Cattle and 13 Buffaloes) will be conserved and developed with the aim of enhancing their productivity and upgrading genetic merit.

Pashu Sanjivni

i) As per Integrated Sample Survey, 90 million cattle & buffaloes are in but their records on breeding, productivity, treatment and vaccination were not properly maintained by the State Animal Husbandry Departments.

ii) System for maintaining records on the animals was not evolved in the country.

iii) Due to absence of records on animal identification and traceability, it was not possible to separate healthy animals from diseased animals and animal products obtained from healthy animals and diseased animals.

iv) Lack of proper animal identification and traceability was major cause of the spread of animal as well as zoonotic diseases in the country.

v) Country was also facing difficulty in expanding trade of milk & milk products, there was no established animal identification and traceability to system to meet sanitary and phytosanitary (SPS) requirements.

e-pashuhaat portal

1. For the first time in the country, under the scheme National Mission on Bovine Productivity, E -Pashudhan Haat portal has been developed for connecting breeders and farmers regarding availability of quality bovine germplasm.
2. Through the portal breeders/farmers can sale or purchase their breeding stock. Information on all forms of germplasm including semen embryos and live animals with all the agencies and stake-holders in the country has been uploaded on the portal. Through this portal, farmers will be aware about the availability of quality disease free bovine germplasm with different agencies in the country.
3. The portal will lead to propagation of high genetic merit germplasm. Through the portal price, evaluation will be available with the farmers/ breeders. Through the portal, there will be no involvement of middlemen in sale and purchase of animals. Portal for sale and purchase of germplasm in all the forms is not available even in the developed dairy countries.
4. This portal will give new dimensions to development and conservation of indigenous breeds as at present information on availability of germplasm of indigenous breeds is not available with the farmers. Information on 7.58 crores semen doses; 371 embryos and 80,039 live animals is available on the portal till date.

Significant initiatives under RGM

1. Awards for encouraging farmers/breeder societies to rear Indigenous breeds of Bovines

- In order to motivate farmers for rearing of indigenous breeds and to enhance the production and productivity of indigenous breeds, the following awards have been instituted under RGM every year:

  a) **Gopal Ratna awards :** For farmers maintaining the best herd of Indigenous Breed(s) and practising best management practices.

  b) **Kamdhenu awards :** For best managed Indigenous Herd by Institutions/Trusts/NGOs/Gaushalas or best managed Breeders' Societies.

Poultry farming

Brooding

Brooding of chicks is very important operation in the early age of the chicks. Chicks are provided with required temperature by artificial means.

Floor Brooding

Spot Brooding

- This conventional method of brooding depends overhang on spot heating either by using electrical bulbs or gas brooders. In this system, one (overhanging or standing) is provided for every 300-600 chicks.
- Hovers are reflectors which provide warmth to chicks using electrical bulbs or gas brooders. The air temperature under the hover is kept at a required level. Hovers are made up of metal, wood or bamboo baskets fitted with electrical bulbs, infra-red bulbs or heating elements and gas heaters.

Brooder Guard

- 2 to 2½ foot away from the hover, brooder guard is provided to prevent chicks straying away from heat source, feed and water. Brooder guards are generally made up of G.I. sheet with foot height. For first 2 days, feed is sprinkled on the paper and after that chick feeders are provided. Chick waterers should be provided first day onwards. Feeders and waterers are arranged in cart wheel manner, so that chick need not have to walk long distance to access feed and water.

Temperature

- The temperature is regulated by adjusting the height of hover have with the number of bulbs/flames. During the first week, brooding temperature under hover (2 inches above litter at the top of the hover) should be 32-34 (90°F-95°F) with a weekly reduction of 5 till it reaches 27 (80°F).

Brooder temperature for chicks at different ages:

Age of Chicks	Brooder Temperature	
	°F	°C
1st Week	90	32.2
2nd Week	85	29.4
3rd Week	80	26.7
4th Week	75	23.9
5th Week	70	21.1
6th Week	70	21.1

Beak Trimming

- Trimming of beak is an important managemental practice. This is done to prevent cannibalism and wastage of feed. Beak trimming is a sensitive operation and it should be done by trained people. The beak trimming is done at 3rd week One third of the beak should be trimmed. There are different methods, cutting and cauterization with hot iron method is popular. Cauterization helps in arresting and destroys the tissue responsible for generating beak growth. Proper care should be taken not to burn the tongue of the chick.
- Use electrolytes and vitamin (K and C) in the water two days before and after beak trimming. Deeper feed should be provided for several days.
- The second beak trimming is done between 12-14 weeks of age.

Duck Farming

- Ducks occupy an important position next to chicken farming in India. They form about 10% of the total poultry population and contribute about 6-7% of total eggs produced in the country. Ducks are mostly concentrated in the Eastern and Southern States of the country mainly coastal region with non-descriptive indigenous stocks, which, however, are poor layers. Central Duck Breeding Farm under Ministry of Agriculture, Government of

India was established in the year 1981 during the 5th Five Year plan with technical collaboration of Government of United Kingdom in order to introduce high yielding variety of ducks for the benefit of farming community.

Important Breeds

1. Khaki Campbell

Khaki Campbell ducks weigh about 2 to 2.2 Kgs, and drakes 2.2 to 2.4 Kgs. Egg size varies from 65 to 75 gms.

2. White Pekin

White Pekin is the most popular duck in the world known for table purpose. It is fast growing and has low feed consumption with fine quality of meat. It attains about 2.2 to 2.5 Kgs of body weight in 42 days of age, with a feed conversion ratio of 1:2.3 to 2.7 Kgs.

- The incubation period of Khaki Campbell duck is 28 days.
- The brooding period of Khaki Campbell ducklings is 3 to 4 weeks.

Important Diseases of Ducks

1. **Duck plague :** Adult birds are mostly affected by virus disease. It is characterized by vascular damage with tissue haemorrhages and free blood in body cavities. The lumina of intestine and gizzard are filled with blood. There is no treatment for the disease. The birds can be protected by Duck Plague Vaccine, available in the country, which is given at the age of 8-12 weeks.
2. **Duck cholera :** It is an infectious disease, caused by bacterial organism Pasteurella Multocoda in ducks over four weeks of age. There is loss of appetite, high body temperature, thirst, diarrhoea and sudden death. Most common lesions are pericarditis, arthritis, petechial and echymotic haemorrhages under the skin (Pink skin), in visceral organs, over the serous surface and intestine (Haemorrhagic enteritis). Liver and spleen are enlarged. The diseases can be controlled by sulpha drugs.
3. **Botulism :** Food poison is a serious problem in both young and adult ducks. It is caused by ingestion of bacterium that grows on decaying plants.

4. **Aflatoxicosis :** It is a condition caused by aflatoxin produced by the mould Aspergillus flavus in the feedstuffs, such as groundnut, maize, rice polish and other tropical feeds on storage. Improper drying of grains, rain and humid weather favour the mould growth. Ducks are very susceptible to aflatoxin content in the feed. Out of the four types of aflatoxins, commonly found viz, B1, B2, G1 and G2. B1 is the most potent toxin. The minimum toxic dose for ducks is 0.03 ppm or 0.03 mg per kg in feed.

Turkey Farming

- Turkey occupies an important position next to chicken, duck. Guinea fowl and quail in contributing the most evolving sector, which is playing a significant role in augmenting the economic and nutritional status of varied population. They form almost two per cent of the total poultry population. They are reared for meat only and its meat is the leanest among other domestic avian species. Turkeys are mostly concentrated in and around cosmopolitan cities of India in small numbers. Indigenous and non-descriptive turkeys are found in good numbers in Kerala, Tamil Nadu, eastern districts of Uttar Pradesh and some other parts of India.
- There are three varieties of turkey commonly available in India. They are broad breasted bronze, Broad breasted white and Beltsville small white. White turkeys seem to be more suitable for Indian conditions.
- Turkeys are not classified into breeds; however, seven standard varieties are available: Bronze, White Holland, Bourbon Red, Narragansett, Black, Slate, Beltsville Small White.
- **Turkey egg :** The turkey will start lay from the 30th week of age and its production period is 24 weeks from the point of lay. Under proper feeding and artificial lightening management, turkey hens lay as much as 60-100 eggs annually. Nearly 70 per cent of the eggs will be laid in the afternoon. The turkey eggs are tinted and weigh about 85 gms. Egg is noticeably pointed at one end with strong shell. The protein, lipid carbohydrate and mineral content of turkey egg are 13.1%, 11.8%,1.7% and 0.8% respectively. The cholesterol is 15.67-23.97 mg/gm of yolk.

- Turkey meat : Turkey meat has nutritional and sensorial properties which make it almost ideal raw material for rational and curative nutrition. People prefer turkey meat because of its leanest nature. The protein, fat, energy value of turkey meat are 24%, 6.6%, 162 Calories per 100 gm of meat. Mineral like potassium, calcium, magnesium, iron, selenium, zinc and sodium are present. It is also rich in essential amino acids and vitamins like niacin, vitamin B6 and B12. It is rich in unsaturated fatty acids and essential fatty acids and low in cholesterol.
- Blue back: Blue back is a permanent dark discolouration of the skin on the back and some-times the side and the breast of the turkeys with dark plumage but not turkeys with white plumage. It is caused by recessive factor.
- Pendulous crop: Pendulous crop is otherwise known as baggy or sour crop. Weakening of the corp. and supporting tissues causes dropped crop so that feed and water accumulate in the organ and pass out slowly or not at all resulting in foul smelling semi-liquid accumulation affecting the crop lining and treatment always useless.

Turkey terminologies

Tom	Adult male turkey
Hem	Adult female turkey
Poult	Young one of turkey
Snood or Dew bill	The fleshy protuberance near the base of the back
Caruncles	The fleshy protuberance on the head and neck usually pink or red in colour which appear from about 5th week of age
Dewlap	A large flap skin seen immediately below the chin
Bread	A tuft of hair attached to the skin of the upper chest region
Strut	Mating behaviour of male turkey
Shooting the red (Similar to an ulcer)	The development of caruncles and this is supposed to indicate the most difficult time in the life of young turkey

Common Diseases of Turkey

Common Diseases of Turkey	Cause	Symptoms	Prevention
Arizonosis	*Salmonella Arizona*	Poults unthrifty and may develop eye opacity and blindness. Susceptible age 3-4 weeks	Elimination of infected breeder flock and hatchery fumigation and sanitation.
Blue comb disease	*Corona virus*	Depression, loss of weight, frothy or watery droppings, darkening of head and skin.	Depopulation and decontamination of farm. Give rest period.
Chronic respiratory disease	*Mycoplasma gallisepticum*	Coughing, gurgling, sneezing, nasal exudates.	Secure *Mycoplasma* free stock
Erysipelas	*Erysipelothrix rhusiopathidae*	Sudden losses, swollen snood, discoloration of parts of face, droppy	Vaccination
Fowl cholera	*Pasturella multocida*	Purplish head, greenish yellow droppings, sudden death	Sanitation and disposal of dead birds.
Fowl pox	*Pox virus*	Small yellow blisters on comb and wattles and scab formation	Vaccination
Haemorrhagic enteritis	*virus*	One or more dead birds	Vaccination
Infectious synovitis	*Mycoplasma gallisepticum*	Enlarged hocks, foot pads, lameness, breast blisters	Purchase clean stock
Infectious sinusitis	*Bacteria*	Nasal discharge, swollen sinuses and coughing	Secure poults from disease free breeders
Mycotoxicosis	*Fungal origin*	Haemorrhages, Pale, fatty liver and kidneys	Avoide feed spoilage
New Castle disease	*Paramyxo Virus*	Gasping, wheezing, twisting of neck, paralysis, soft shelled eggs	Vaccination
Paratyphoid	*Salmonella pullorum*	Diarrhea in poults	Prevention and flock sanitation

Sources:

- 19th Livestock Census (2014). DADF, MoA, GoI
- APEDA Market Report (2017). Global Agriculture Information Network, US Foreign Agriculture Service.
- Annual Report (2017). DADF, MoA, GoI
- Basic Animal Husbandry & Fisheries Statistics (2017). Animal Husbandry Statistics Division, DADF, MoA, GoI

❑❑❑

Chapter 9

Fisheries

Indian Fisheries

- With diverse resources ranging from deep seas to lakes in the mountains and more than 10% of the global biodiversity in terms of fish and shellfish species, the country has shown continuous and sustained increments in fish production since independence. Constituting about 6.3% of the global fish production, the sector contributes to 1.1% of the GDP and 5.15% of the agricultural GDP.
- The total fish production of 10.07 million metric tonnes presently has nearly 65% contribution from the inland sector and nearly the same from culture fisheries.

The National Fisheries Development Board (NFDB) was established in **2006** as an autonomous organization under the administrative control of the Department of Animal Husbandry, Dairying & Fisheries, Ministry of Agriculture and Farmers Welfare, Government of India to enhance fish production and productivity in the country.

Global position	3rd in Fisheries 2nd in Aquaculture
Contribution of Fisheries to GDP (%)	1.07
Contribution to Agri. GDP (%)	5.15
Per capita fish availability (Kg.)	9.0
Annual Export earnings (Rs. in Crore)	33,441.61
Employment in sector (million)	14.0

Fish Production Data

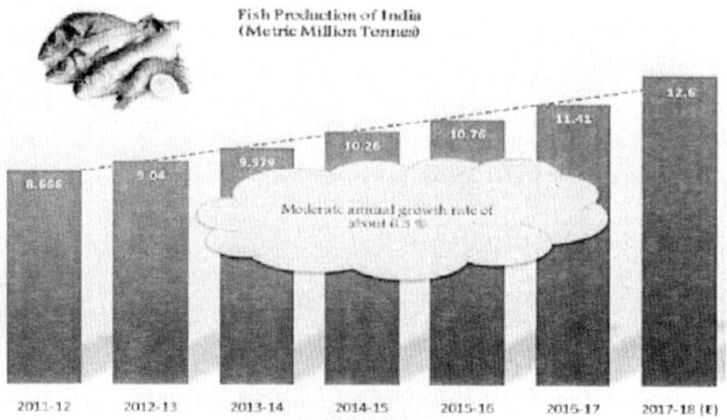

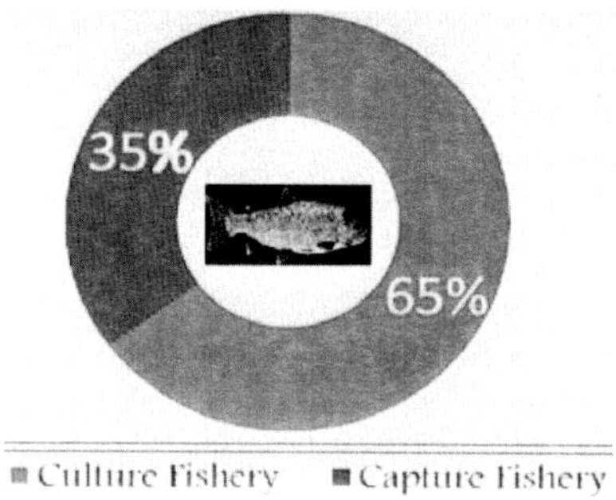

'State Fish'- declared by some of the States with a Concept for the Conservation of Regionally Important **& Threatened Fish Species:**

1. West Bengal - ***Tenualosa ilisha***
2. Uttar Pradesh **-** ***Chitala chitala***
3. J&K, H.P, UK, Arunachal Pradesh **-** ***Tor putitora***
4. Kerala **-** ***Etroplus suratensis***
5. Bihar **-** ***Clarias batrachus (Magur)***
6. Odisha **-** ***Tor mahanadicus***
7. Manipur **-** ***Osteobrama belangeri***

Blue –Revolution: Neel Kranti Mission

The Neel Kranti Mission, 2016 (NKM-16), being the year in which it was envisioned by Hon'ble Prime Minister, will have a multi-dimensional approach to all activities concerned with development of fisheries sector.

Objectives of Neel Kranti Mission

- To fully tap the total fish potential of the country, both in the inland and the marine sector and triple the production by 2020.
- To transform the fisheries sector as a modern industry with special focus on new technologies and processes.
- To double the income of the fishers and fish farmers with special focus on increasing productivity and better marketing post-harvest infrastructure including e-commerce and other technologies and global best innovations.
- To ensure inclusive participation of the fishers and fish farmers in the income enhancement.
- To triple the export earnings by 2020 with focus on benefits flow to the fishers and fish farmers including through institutional mechanisms in the co-operative, producer companies and other structures.
- To enhance food and nutritional security of the country.

Central sponsored scheme on blue revolution pattern of financial assistance

Beneficiary oriented projects funded by DADF, Govt. of India.

A) Approved project cost share

Category	Govt. assistance	Beneficiaries share	Total
General Category	40%	60%	100%
SC/ST/ Women and their co-operatives	60%	40%	100%

B) Central and state share of Govt. assistance

Region	Central share	State share	Total
Other state	60%	40%	100%
North-East and hilly state	90%	10%	100%
Union territories	100%	0%	100%

Farming of Tilapia in India

Tilapia is native to Africa and Middle East, and has emerged from mere obscurity to one of the most productive and internationally traded food fish in the world. The farming of tilapias, especially of Nile tilapia (*Oreochromis niloticus*) in its crudest form is believed to have originated more than 4000 years ago from Egypt. The first recorded scientifically oriented culture of tilapia was conducted in Kenya in 1924 and soon spread to many parts of the world.

Species for aquaculture

Tilapia belongs to the family Cichlidae under order Perciformes. The tilapias have recently been classified into three genera, based on parental incubation of eggs. The species of the genera *Sarotherodon* and *Oreochromis* are mouth brooders, while *Tilapia* incubates eggs in a lake or pond bottom built-in "nest". There are about 70 species of tilapias, of which nine species are used in aquaculture worldwide (FAO, 2008). Important commercial species include: the Mozambique tilapia (*Oreochromis mossambicus*), Blue tilapia (*O. aureus*), Nile tilapia (*O. niloticus*), Zanzibar tilapia (*O. hornorum*), and the Red belly tilapia (*O. zilli*).

Habitat and Biology

Nile tilapia is a tropical species that prefers to live in shallow water. It is an omnivorous grazer that feeds on phytoplankton, periphyton, aquatic plants, small invertebrates, benthic fauna, detritus and bacterial films associated with detritus. Sexual maturity in ponds is reached at an age of 5-6 months. Spawning begins, when the water temperature reaches 24°C. The female incubates the eggs in her mouth and broods the fry after hatching until the yolk sac is absorbed. Fecundity is proportional to the body weight of the female. A 100 g female will produce about 100 eggs per spawn, while a female weighing 600-1000 g can produce 1000 to 1500 eggs. Nile tilapia can live longer than 10 years and reach a weight exceeding 5 kg.

In India, tilapia (*Oreochromis mossambicus*) was introduced in 1952, with a view to filling up unoccupied niches, such as ponds and reservoirs. The species spread all across the country within a few years due to its prolific breeding and adaptability to wide range of environmental condition.

The Nile tilapia was introduced to India during late 1970s. In 2005, River Yamuna harboured only negligible quantity of Nile tilapia, but in two years' time, its proportion has increased to about 3.5% of total fish species in the river. Presently in the Ganges river system, proportion of tilapia is about 7% of the total fish species.

World Fisheries Day: 21 November

Freshwater Aquaculture

The available freshwater aquaculture technologies can be classified into technologies for production of table size fish or grow out and fish seed production. Both types of technologies are for different categories of fishes, i.e. carps and catfishes including air-breathing fishes, prawn and ornamental fishes. Their catalogue is mentioned below:

The current freshwater grows out technologies may be classified into

i) Polyculture of Indian carps or Indian and exotic carps together (Composite carp culture)

ii) Mono- and polyculture of air-breathing fishes

iii) Mono- and polyculture of freshwater prawns, and

iv) Integrated fish farming.

The composite carp culture can further be classified into:

a) Low input or fertilizer-based system;

b) Medium input or fertilizer- and feed-based system;

c) High input or intensive feed and aeration-based system;

d) Sewage fed water-based system; and

e) Aquatic weed-based system.

Integrated fish farming includes:

a) Paddy-cumfish culture,

b) Fish-cum-cattle farming,

c) Pig-cum-fish farming,

d) Duckcum-fish culture, and

e) Poultry-cum-fish farming

The technologies for fish breeding and seed production may be categorized as:

i) Induced breeding of carps and strain development

ii) Intensive carp seed rearing

iii) Breeding and seed production of air breathing catfishes

iv) Breeding and seed production of giant freshwater prawn, and

v) Ornamental fish

Brackishwater Aquaculture

The grow out shrimp farming is the most important brackishwater aquaculture technology in India.

The other technologies include:

i) Mud crab fattening

ii) Edible oyster farming,

iii) Mussel farming,

iv) Clam culture, and

v) Finfish farming.

Hatchery for seed production technologies include:

i) Bivalve hatchery of mussel, clam and edible oyster, and

ii) Shrimp and sea-bass hatchery. Hatchery input technologies used were live feed, micro-algae, rotifer and brine shrimp (Artemia) culture.

Mariculture

This includes technologies for pearl oyster farming and pearl production, mussel, sea cucumber and seaweed. Hatchery and seed production technologies have been developed for pearl oyster, ornamental fish, clown fish, damsel fish and sea horse.

Cultivable Fish Species

Cultivable fish species can be categorized into three types:

1. Indigenous or the native fresh water fishes. Exm. - Major Carps
2. Salt water fishes that are acclimatized to fresh water. Exm. - Chanos and Mullets
3. Exotic fish which is imported from outside of countery. Exam. - Mirror carp, Chinese carp, Tilapia, etc.

Cultivable Species of Fishes

In India, mostly, major carps are used to cultivable fish and some catfish also use to culture

- **The important cultivable species are:**

  Labeo rohita, Catla catla, Cirrhinus mrigala, Labeo calbasu, L. bata **and some exotic major carps species are:** *Cyprinus carpio, Ctenopharyngodon idella, Hypothalmichthys molitrix, Tilapia mossambica,* **etc.**

- **And some cat fish species are:**

  Clarias batrachus, Heteropneustes fossilis, Wallago attu, **etc.**

- **Some other species are:**

  Anabas testudineus, Channa sps., Salmo sps., **etc.**

Important points:

- Stocking of fishes of different habits, in same pond is called Composite fish culture.
- Soil quality required for stocking - Every 100 gm should contain 50-75 mg nitrogen, 6-12 mg phosphorus and 1.5-2.5 gm organic carbon.
- The tanks, which are **acidic in nature,** are **less productive than alkaline ponds.**
- Lime is **used to bring the pH to the desired level.**
- The **addition of lime** also hastens **organic decomposition.**
- The normal doses of the lime desired ranges from 200 to 250 Kg/ha.

- In composite fish culture in stocking pond, fish seed of **10-15 cm length** (fingerling) is stocked at the rate of **5000-7000 nos/hectare.**
- Water Temperature required 20°C to 37°C good for fish farming.
- pH of pond water 6.7 to 8.6; Phytoplankton require a pH of about 7 and zooplankton a slightly lower pH of 6.5.
- Indian major carps can tolerate turbidity upto 2000 ppm.
- Dissolved Oxygen content required for Carps need 6-7 mg/litre dissolved oxygen for aquatic breathing.
- Harvesting time-Approximate number of months required to reach the marketable size: 7-8 months.
- For good production from cultured fish – Soil pH should be between 7.5 to 8.5.
- Area required Nursery pond 100-500 m^2; depth 1-1.5 m; Fish reared upto fry stage for 2-3 weeks.
- Area required Rearing pond 500-1000 m^2; depth 1.5-2 m; Fish reared upto fingerling stage for about 3 months.
- Area required Stocking pond 1000-2000 m^2; depth 2-2.5 m; Fish reared upto Fishes are stocked until reaching the market size.

Lime requirement depend on Soil pH

Soil pH	Lime (kg/ha)
4.5-5.0	2000
5.1-6.5	1000
6.6-7.5	500
7.6-8.5	200
8.6-9.5	Nil

Important Questions on FISHERIES

1. *Tilapai fish is also Known as:*

   a) Kowai b) Ari

   c) Rohu d) Black mahseer

2. *Scientific name of Mud carp:*

   a) Labeo calbasu b) Labeo rohita

   c) Cirrhina molitorella d) Cirrhina mrigla

3. *Chanos Chanos is:*

   a) Milk fish b) Tilapia

   c) Big gourami d) Black carp

4. *The capacity of an individual fish to produce female reproductive cells is known as:*

   a) Fecundity b) Re-stocking

   c) Embankment d) Fish reproduction

5. *Area of Nursery pond is.........*

   a) 100-500 m^2 b) 500-1000 m^2

   c) 1000-2000m^2 d) 50 m^2

6. *In rearing pond, depth of water ranges from:*

   a) 1.5-2.0 m b) 2.5-5.0 m

   c) 10.5-11.5m d) 1.0 m

7. *Salmo is related to:*

   a) Trout b) Magur

   c) Tilapia d) Sarana

8. *Which is classified under weed fish group:*

   a) Clupisoma b) Ailia

   c) Channa **d) Chela**

9. *The branch of science dealing with the study of fishes:*

   a) Ichthyology b) Aquaculture

   c) Limnology d) Hydrology

10. *Pond covers 5% area of total productive area of fish farming:*

    a) Nursery pond b) Rearing pond

    c) Stocking pond d) Re-stocking pond

11. *Pond suitable for undulating and hilly areas:*

    a) Embankment pond b) Dug out pond

    c) Hilly pond d) Rearing pond

12. *Central institute of freshwater aquaculture situated at:*
    a) Bhopal b) Kerala
    c) Bhubaneshwar d) Tamil Nadu
13. *Central institute of fisheries education situated at:*
    a) Mumbai b) Delhi
    c) Kerala d) Tripura
14. *Central institute of fisheries education establish in the year:*
    a) 1961 b) 1956
    d) 1976 c) 1991
15. *Matsya Darpan publish by:*
    a) Central institute of fisheries education
    b) ICAR
    c) Central institute of freshwater aquaculture
    d) IARI
16. *Cod liver oil is rich in:*
    a) Vitamins A and D b) Vitamins A and C
    c) Vitamins A and B d) Vitamins A and E
17. *Crustacean fishery is connected with exploitation of :*
    a) Oysters and Crabs b) Mussels and Squids
    c) Shells and Cuttle fish **d) Lobster and Prawn**
18. *Cultivation of fishes in artificially prepared ponds or water bodies is called:*
    a) Aquaculture **b) Pisciculture**
    c) Vermiculture d) Agriculture
19. *Drying of fishes by lowering temperature is:*
    a) Smoking **b) Freeze drying**
    c) Salting d) Chilling
20. *Fish flour is rich in:*
    a) Fat **b) Protein**
    c) Vitamins d) Minerals

21. *Fish introduced in India by foreigners is:*

a) Labeo rohita | b) Mystus singhala
c) Pomfret | d) Clarias batrachus

22. *Fishes reared in culture fishery in India are:*

a) Salmon and Rohu | b) Salmon and Catla
c) Catla ad Magur | **d) Rohu and Catla**

□□□

Chapter 10

Apiculture

Beekeeping in India

In India, first attempt to keep bees in movable frame hives was made in 1882, in Bengal and then in 1883-84 in Punjab.

It is important to know the difference between a species and sub-species. Species are reproductively isolated from each other and these cannot interbreed, whereas sub-species are geographically isolated and can interbreed **Among the two domestic bee species, each has many sub-species in different parts of the world e.g. *Apiscerana* has three sub-species in India:**

- A. *ceranacerana* in Himachal Pradesh and Jammu and Kashmir (North India)
- A. *ceranaindica* in Kerala, Tamil Nadu and Karnataka. (South India)
- A. *ceranahimalaya* in Nagaland, Manipur, Mizoram, Assam and Meghalaya. (Eastern parts of India)

In addition to above three sub-species, *A. cerana japonica* has been identified from Japan:

Nesting	Apisdorsata	Apisflorea	Apiscerana	Apismellifera
	Open nesting. Builds single large comb (ca 1m2) attached to branches of trees or rocks, etc.	Open nesting. Builds single small comb (ca size of palm of hand) fixed to branches of bushes.	Cavity nesting. Builds many parallel combs in cavities of tree trunks, hollows of rocks, poles and other covered places	Cavity nesting and similar in habits to Apiscerana and builds parallel combs.
Distribution in India	Found in plains as well as hills up to 1600 metres above sea level. Highly migratory.	Found in plains up to 300 metres above sea level. Highly migratory.	Found throughout India having 3 sub-species	Exotic bee to India. Introduced successfully in 1962. It has many sub-species (more than 23)throughout world

Size	Biggest honey bee (16-18mm)	Smallest Apis bee (9-10mm)	Medium size (14-15mm)	Medium size (14-16mm)
Swarming/ Absconding	Strong tendency	Strong tendency	Strong tendency	Only in African sub-species
Temperament	Furious	Mild	Furious	Gentle except African sub-species
Average honey yield per colony/year	40 kg (wild bees; cannot be domesticated)	500 g (wild bees; cannot be domesticated)	5 kg (Hive bees; can be domesticated)	15 kg Hive bees; can be domesticated)
Method of honey extraction	By squeezing (unhygienic)	By squeezing (unhygienic)	By centrifugal honey extractor from the hived bees (hygienic).	By centrifugal honey extractor from the hived bees (hygienic).
Number of cells/10cm comb (worker cells)	18-19	32-36	21-25	17-19

Development: The developmental stages of honeybees are: egg, larva, pupa and the adult.

Duration of life stages of different castes of honeybees varies which is given in the table

Caste	Egg period (days)		Larval Stage(days)		Pupal Stage (days)		Total (days)	
	A. cerana	A. mellifera	A. cerana	A. mellifera	A. cerana	A. mellifera	A. cerana	A. mellifera
Queen	3	3	5	5	7-8	8	15-16	16
Workers	3	3	5-7	5	11-12	12-13	18-20	21
Drone	3	3	7	7	14	14	24	24

Important term in apiculture

Apiculture - Beekeeping or Apiculture is the maintenance and raising of Honeybee colonies by a beekeeper.

Apiology - The study of honey bees is known as Apiology.

What is swarming?

This is a natural instinct for increase in the number of colonies. Division of colony takes place in which worker bees (30 to 70 per cent), fill their honey stomachs with the food and leave the colony along with old queen and this divide, called as swarm, settles down temporarily generally in the nearby area of the colony on the bushes, hedges, tree branches, etc.

Time of swarming

Time to issue swarms by the colonies is from 10 a.m. to 2 p.m. on sunny days. If weather is not favourable, swarms may be issued even earlier in the morning or late in the evening.

What is honey flow?

It is the period when honeybees gather and store surplus honey in the hive after attaining peak population in the colony.

A **honey super** is a part of a commercial or other managed (such as by a hobbyist) beehive that is used to collect honey. The most common variety is the "Illinois" or "medium" super with a depth of $6^{5/8}$ inches, in the length and width dimensions of a Langstroth hive.

Drone

Drones are male bees and have no sting. They are broader built than worker bees or the queen and have much bigger eyes. Drones don't perform any colony duties and need to be fed by worker bees. Their main purpose is impregnating a virgin queen.

Castes -The three types of bees in the colony: queen, drones and worker bees.

Swarming season

The time of year when bee swarms usually emerge; usually spring to early summer, September to December.

Thorax

The central region of an insect to which the wings and legs are attached.

Requeen

To replace an existing queen by removing her and introducing a new queen.

Royal jelly

Royal jelly is a protein-rich, milky white secretion of the hypopharyngeal gland of nurse bees, used to feed the queen, all young bee larvae for the first two days of their existence and the queen larva until it pupates.

Queen

The queen bee is the only reproductive female in the colony. Her head and thorax are similar in size to that of the worker but has a longer and plumper abdomen than a worker. The queen also has a stinger but its barbs are reduced. That's why, she does not die when she uses it.

Nurse bees

Young bees, usually three to ten days old. Nurse bees feed and take care of developing brood, feed the drones and the queen.

Pollen

Pollen is also collected by the bees from flowers and stored in comb cells as food reserve. Pollen is the protein in the bees' diet.

❑❑❑

Chapter 11

Minimum Support Prices Recommended by CACP and Fixed by Government, 17-18

Cabinet approves increase in MSP for Rabi crop season, 2018-19

The Cabinet, in its meeting held on October 03, 2018, increased the minimum support price (MSP) of various rabi crops with the onset of the rabi crop season (2018-19). The Cabinet, in its meeting held on October 03, 2018, increased the minimum support price (MSP) of various rabi crops with the onset of the rabi crop season (2018-19). The price hikes are in the range of 2-24% from previous year's MSP prices for all major rabi crops, as per media sources.

MSP for Rabi Crop Season, 2018-19

Rabi Crops (Rs per quintal)	Cost of Production	MSP in 2017-18 season	MSP in 2018-19 season	Hike in MSP (%)
Wheat	866	1,735	1,840	6.1
Barley	860	1,410	1,440	2.1
Gram	2,637	4,250	4,620	8.7
Masur (Lentil)	2,532	4,150	4,475	7.8
Mustard	2,212	3,900	4,200	7.7
Safflower	3,294	4,000	4,945	23.6

Commodity	Variety	MSP for 2017-18 Season	MSP Approved for 2018-19 Season	Percentage increase
Rice	Common	1550	1750	12.90
	Grade A Grade A	1590	1770	11.32
Jowar	Hybrid	1700	2430	42.94
	Maldandi Maldandi	1725	2450	42.03
Bajra	-	1425	1950	36.84
Ragi	-	1900	2897	52.47
Maize	-	1425	1700	19.30
Arhar (Tur)	-	5450	5675	4.03
Groundnut	-	4450	4890	25.11
Moong	-	5575	6975	3.70
Urad	-	5400	5600	9.89
Sunflower Seed	-	4100	5388	31.42
Soyabean	-	3050	3399	11.44
Sesamum	-	5300	6249	17.91
Nigerseed	-	4050	5877	45.11
Cotton	Medium Staple	4020	5150	28.11
	Long Staple	4320	5450	26.16

❑❑❑

Chapter 12

State Agricultural Universities and Institutes

State Agricultural Universities

S. No.	Name of University
	Andhra Pradesh
1	Acharya NG Ranga Agricultural University, Guntur
2	Dr. YSRHU (APHU), Venkataramannagudem
3	Sri Venkateshwara Veterinary University, Tirupati
	Assam
4	Assam Agricultural University, Jorhat
	Bihar
5	Bihar Agricultural University, Sabour, Bhagalpur
6	Bihar Animal Sciences University, Patna
	Chhattisgarh
7	Indira Gandhi Krishi Vishwavidyalaya, Raipur
8	Chhattisgarh Kamdhenu Vishvavidyalaya, Durg
	Gujarat
9	Sardar Krushinagar Dantiwada Agricultural University, Dantiwada
10	Anand Agricultural University, Anand
11	Navsari Agricultural University, Navsari
12	Junagarh Agricultural University, Junagarh
13	Kamdhenu University, Gandhinagar
	Haryana
14	Chaudhary Charan Singh Haryana Agricultural University, Hisar
15	Lala Lajpat Rai University of Veterinary & Animal Sciences, Hisar
16	Haryana State University of Horticultural Sciences, Karnal
	Himachal Pradesh
17	Ch. Shrawan Kumar Himachal Pradesh Krishi Vishwavidyalaya, Palampur

18	**Dr. Yashwant Singh Parmar University of Horticulture & Forestry, Solan**
	Jharkhand
19	Birsa Agricultural University, Ranchi
	Jammu & Kashmir
20	Sher-e-Kashmir University of Agricultural Science & Technology, Srinagar
21	Sher-e-Kashmir University of Agricultural Science & Technology, Jammu
	Karnataka
22	University of Agricultural Sciences, Bangalore
23	Karnataka Veterinary, Animal and Fisheries Sciences University, Bidar
24	University of Agricultural Sciences, Raichur
25	University of Agricultural Sciences, Dharwad
26	University of Horticulture Science, Bagalkot
27	University of Agriculture & Horticulture Sciences, Shimoga
	Kerala
28	Kerala Agricultural University, Thrissur
29	Kerala University of Fisheries and Ocean Studies, Panangad, Kochi
30	Kerala Veterinary and Animal Sciences University, Pookode, Wayanand, Kerala
	Madhya Pradesh
31	Rajmata Vijayaraje Scindia Krishi Vishwavidyalaya, Gwalior
32	Nanaji Deshmukh Pashu Chikitsa Vishwavidyalaya, Jabalpur
33	Jawaharlal Nehru Krishi Vishwavidyalaya, Jabalpur
34	**Dr. Balasahib Sawant Kokan KrishiVidyapeeth, Dapoli**
35	Maharastra Animal & Fisheries. Sciences University, Nagpur
36	Vasantrao Naik Marathwada Krishi Vidyapeeth, Parbhani
37	Matatam Phule Krishi Vidyapeeth, Rahuri
38	Dr. Punjabrao Deshmukh Krishi Vishwavidyalaya, Akola
	Orissa
39	Orissa University of Agricultural & Technology, Bhubaneshwar
	Punjab
40	Guru Angad Dev Veterinary and Animal Sciences University, Ludhiana
41	Punjab Agricultural University, Ludhiana

	Rajasthan
42	Maharana Pratap University of Agriculture & Technology, Udaipur
43	Swami Keshwanand Rajasthan Agricultural University, Bikaner
44	Rajasthan University of Veterinary & Animal Sciences, Bikaner
45	SKN Agriculture University, Jobner
46	Agriculture University, Kota
47	Agriculture University, Jodhpur
	Tamil Nadu
48	Tamil Nadu Agricultural University, Coimbatore
49	Tamil Nadu Veterinary & Animal Sciences University, Chennai
50	Tamil Nadu Fisheries University, Nagapattanam
	Telangana
51	Sri Konda Laxman Telangana State Horticultural University, Hyderabad
52	Sri PV Narsimha Rao Telangana Veterinary University, Hyderabad
53	Professor Jayashankar Telangana State Agricultural University, Hyderabad
	Uttarakhand
54	G.B. Pant University of Agriculture & Technology, Pantnagar
55	VCSG Uttarakhand University of Horticulture & Forestry, Bharsar
	Uttar Pradesh
56	Chandrashekhar Azad University of Agricultural & Technology, Kanpur
57	Narendra Deva University of Agriculture & Technology, Faizabad
58	Sardar Vallabhbhai Patel University of Agriculture & Technology, Meerut
59	U.P. Pt. Deen Dayal Upadhyaya Pashu Chikitsa Vigyan Vishwavidhyalaya Evam Go Anusandhan Sansthan, Mathura
60	Banda University of Agricultural and Technology, Banda
61	Sam Higginbottom University of Agriculture, Technology & Sciences, Allahabad
	West Bengal
62	Bidhan Chandra Krishi Vishwavidhyalaya, Mohanpur
63	West Bengal University of Animal & Fishery Sciences, Kolkata

Deemed Universities

1. ICAR-Indian Agricultural Research Institute, New Delhi
2. ICAR-National Dairy Research Institute, Karnal

3. ICAR-Indian Veterinary Research Institute, Izzatnagar
4. ICAR-Central Institute on Fisheries Education, Mumbai

Central Agricultural Universities

1. Central Agricultural University, P.O. Box 23, Imphal-795004, Manipur
2. Rani Laxmi Bai Central Agricultural University, Jhansi, Uttar Pradesh
3. Dr. Rajendra Prasad Central Agricultural University, Pusa (Samastipur)

National Research Centres - 15	
1.	ICAR-National Research Centre for Banana, Trichi
2.	ICAR-National Research Centre for Grapes, Pune
3.	ICAR-National Research Centre for Litchi, Muzaffarpur
4.	ICAR-National Research Centre for Pomegranate, Sholapur
5.	ICAR-National Research Centre on Camel, Bikaner
6.	ICAR-National Research Centre on Equines, Hisar
7.	ICAR-National Research Centre on Meat, Hyderabad
8..	ICAR-National Research Centre on Mithun, Medziphema, Nagaland
9.	ICAR-National Research Centre on Orchids, Pakyong, Sikkim
10.	ICAR-National Research Centre on Pig, Guwahati
11.	ICAR-National Research Centre on Plant Biotechnology, New Delhi
12.	ICAR-National Research Centre on Seed Spices, Ajmer
13.	ICAR-National Research Centre on Yak, West Kemang
14.	ICAR-National Centre for Integrated Pest Management, New Delhi
15.	National Research Centre on Integrated Farming (ICAR-NRCIF), Motihari

National Bureaux - 6

1. National Bureau of Agricultural Insect Resources, Bengaluru -1957
2. National Bureau of Agriculturally Important Micro-Organisms, Mau Nath Bhanjan, Uttar Pradesh -2001

3. National Bureau of Animal Genetic Resources, Karnal, Haryana -1984
4. National Bureau of Fish Genetic Resources, Lucknow, Uttar Pradesh -1983
5. National Bureau of Plant Genetic Resources, New Delhi -1976
6. National Bureau of Soil Survey & Land Use Planning, Nagpur, Maharashtra -1986

Directorates/Project Directorates -13

1.	ICAR-Directorate of Groundnut Research, Junagarh
2.	ICAR-Directorate of Soyabean Research, Indore
3.	ICAR-Directorate of Rapeseed & Mustard Research, Bharatpur
4.	ICAR-Directorate of Mushroom Research, Solan
5.	ICAR-Directorate on Onion and Garlic Research, Pune
6.	ICAR-Directorate of Cashew Research, Puttur
7..	ICAR-Directorate of Medicinal and Aromatic Plants Research, Anand
8.	ICAR-Directorate of Floricultural Research, Pune, Maharashtra
9.	ICAR-Directorate of Weed Research, Jabalpur
10.	ICAR-Project Directorate on Foot & Mouth Disease, Mukteshwar
11.	ICAR-Directorate of Poultry Research, Hyderabad
12.	ICAR-Directorate of Knowledge Management in Agriculture (DKMA), New Delhi
13.	ICAR-Directorate of Cold-Water Fisheries Research, Bhimtal, Nainital

Fisheries Institutes under the control of the Department

1. Central Institute of Fisheries, Nautical & Engineering Training (CIFNET), Cochin
2. Central Institute of Coastal Engineering for Fishery, (CICEF), Bangaluru
3. Fishery Survey of India (FSI), Mumbai
4. National Institute of Fisheries Post Harvest Technology and Training (NIFPHATT), Cochin
5. Coastal Aquaculture Authority (CAA), Chennai
6. National Fisheries Development Board (NFDB), Hyderabad
7. National Federation of Fishers Co-operatives Ltd. (FISHCOPFED), New Delhi

❑❑❑

Chapter 13

Chief/Director of Institutes of National and International Bodies

Chief/Director of Institutes of National and International Bodies

1. ICAR: DG- Dr. T. Mahapatra
2. ASRB: Dr. Gurubachan
3. UN Secretary: António Manuel de Oliveira Guterres
4. UN Members: 193 countries
5. International Monetary Fund (IMF) New Chief: Christine Lagarde
6. DG of FAO: Jose Graziano Da sliva
7. Chairman of NABARD: Dr. Harish Kumar Bhanwala
8. Governor of RBI: Urjit Patel
9. President of World Bank: Jim Yong Kim
10. World Bank Chief Economist : Pinelopi Koujianou Goldberg
11. NITI (National Institution for Transforming India) Aayog CEO: Amitabh Kant
12. NITI Aayog Chairman: Narendra
13. NITI Aayog Vice Chairman: Dr. Rajiv Kumar
14. Director of IMD: K J Ramesh
15. Chairman of National Biodiversity Authority (NBA) : Dr. B. Meena Kumari
16. Secretary General of World Meteorological Organization : Petteri Taalas
17. Deputy Chairman, Planning Commissions- Montek Singh Ahluwalia.

18. Chairman of National Commission for Farmers- Dr. M.S. Swaminathan
19. Union Minister of Agriculture and Farmer Welfare is the ex-office President of the ICAR Society (President- Radha Mohan Singh).
20. Secretary, Department of Agricultural Research & Education Ministry of Agriculture, Govt. of India & Director General, ICAR-the Principal Executive Officer of the Council. (Present DG– Dr. T. Mahapatra)
21. Agricultural Scientists Recruitment Board ,Chairman-Dr. Guru Bachan Singh (Agronomist)

Institutes and their Establishing Year

1. Central Agro-forestry Research Institute, Jhansi -1988
2. Central Arid Zone Research Institute, Jodhpur -1952
3. Central Avian Research Institute, Izzatnagar -1927
4. Central Coastal Agricultural Research Institute, Ela, Old Goa, Goa -1976
5. Central Inland Fisheries Research Institute, Barrackpore -1947
6. Central Institute of Brackishwater Aquaculture, Chennai -1987
7. Central Institute for Arid Horticulture, Bikaner -1976
8. Central Institute for Research on Buffaloes, Hisar -1985
9. Central Institute for Research on Cattle, Meerut, Uttar Pradesh -1987
10. Central Institute for Research on Goats, Makhdoom -1979
11. Central Institute for Women in Agriculture, Bhubaneshwar -1996
12. Central Institute of Agricultural Engineering, Bhopal -1976
13. Central Institute of Cotton Research, Nagpur -1976
14. Central Institute of Fisheries Technology, Cochin -1954
15. Central Institute of Freshwater Aquaculture, Bhubneshwar -1987
16. Central Institute of Research on Cotton Technology, Mumbai -1924
17. Central Institute of Sub-Tropical Horticulture, Lucknow -1972

18. Central Institute of Temperate Horticulture, Srinagar - 1994
19. Central Institute of Post-Harvest Engineering and Technology, Ludhiana -1989
20. Central land Agricultural Research Institute, Port Blair –1978
21. Central Marine Fisheries Research Institute, Kochi -1947
22. Central Plantation Crops Research Institute, Kasargod -1970
23. Central Potato Research Institute, Shimla -1949
24. Central Research Institute for Jute and Allied Fibers, Barrackpore-1938
25. Central Research Institute of Dryland Agriculture, Hyderabad - 1985
26. Central Sheep and Wool Research Institute, Avikanagar, Rajasthan -1962
27. Central Soil Salinity Research Institute, Karnal -1969
28. Central Tobacco Research Institute, Rajahmundry -1947
29. Central Tuber Crops Research Institute, Trivandrum -1963
30. ICAR Research Complex for Eastern Region, Patna -2001
31. ICAR Research Complex for NEH Region, Barapani
32. Indian Agricultural Research Institute, Jharkhand
33. Indian Agricultural Statistics Research Institute, New Delhi -1930
34. Indian Grassland and Fodder Research Institute, Jhansi -1962
35. Indian Institute of Agricultural Bio-technology, Ranchi -2014
36. Indian Institute of Farming Systems Research, Modipuram - 1952-53
37. Indian Institute of Horticultural Research, Bengaluru -1967 and shifted -1968
38. Indian Institute of Maize Research, New Delhi
39. Indian Institute of Millets Research, Hyderabad -1958
40. Indian Institute of Natural Resins and Gums, Ranchi -1920 renamed, 2007
41. Indian Institute of Oil Palm Research, Eluru -1995 and upgraded -2014

42. Indian Institute of Oilseeds Research, Hyderabad -1977
43. Indian Institute of Pulses Research, Kanpur -1983
44. Indian Institute of Rice Research, Hyderabad project -1965 and upgraded -2014
45. Indian Institute of Seed Research, Mau – Project -1979 Directorate of Seed Research started on-2004
46. Indian Institute of Soil Sciences, Bhopal -1988
47. Indian Institute of Soil and Water Conservation, Dehradun -1974
48. Indian Institute of Spices Research, Calicut -1976
49. Indian Institute of Sugarcane Research, Lucknow -1952
50. Indian Institute of Vegetable Research, Varanasi -1999
51. Indian Institute of Water Management, Bhubaneshwar -1988
52. Indian Institute of Wheat and Barley Research, Karnal
53. National Academy of Agricultural Research & Management, Hyderabad -1976
54. National Institute of Abiotic Stress Management, Malegaon, Maharashtra -2009
55. National Institute of Agricultural Economics and Policy Research, New Delhi -1991
56. National Institute of Animal Nutrition and Physiology, Bengaluru -1995
57. National Institute of Biotic Stresses Management, Raipur -1912
58. National Institute of High-Security Animal Diseases, Bhopal -2014
59. National Institute of Research on Jute and Allied Fibre Technology, Kolkata -1938
60. National Institute of Veterinary Epidemiology and Disease Informatics, Hebbal, Bengaluru -
61. National Rice Research Institute, Cuttack -1946
62. Sugarcane Breeding Institute, Coimbatore -1912
63. Vivekananda Parvatiya Krishi Anusandhan Sansthan, Almora -1924

❑❑❑

Chapter 14

Points to be Remembered

Indian Agricultural Research Journals

S.No.	Name of Journal	Published by	Period
1	Indian Journal of Agricultural Sciences	Indian Council of Agricultural Research	Monthly
2	Indian Journal of Agronomy	Indian Society of Agronomy, Indian Agricultural Research Institute, Pusa	Quarterly
3	Indian Farming	Indian Council of Agricultural Research	Monthly
4	Progressive Farming	Punjab Agricultural University, Ludhiana, India	Monthly
5	Indian Journal of Horticulture	Horticulture Society of India, Bengaluru	Quarterly
6	Agriculture Today	Center for Agriculture and Rural Development, New Delhi	Monthly
7	Kheti	The Business Manager, Agricultural Research Communication Center, Karnal	Monthly
8	Modern Kethi (Hindi)	Mehram Publication	Bi-weekly
9	Phal Phool (Hindi)	ICAR	Bi-monthly
10	Indian Horticulture	ICAR	Bi-monthly
11	Baliraja	Agro India	Monthly
12	Farm N Food (Hindi)	Delhi press	Weekly
13	Bhoomi	K.N.K. foundation	Bi-monthly

14	Agricultural Situation in India	Director of Economic & Statistics, Min. of Agriculture	Monthly
15	Agricultural Marketing	Deptt. of Agriculture Cooperation, Min. of Agriculture, Dte. of Mktg. & Inspection, Faridabad	Quarterly
16	Fertiliser Marketing News	Fertilizer Association of India	Monthly
17	Indian Journal of Fertilisers	Fertilizer Association of India	Monthly
18	Khad Patrika (Hindi)	Fertilizer Assiciation of India	Monthly
19	Journal of Agro-meteorology	Association of Agro-meteorlogists Anand Agricultural University, Anand	Six-monthly
20	Mausam	India Meteorological Department, New Delhi	Quarterly

Agriculture Related Years

1. 2004 : International Year of Rice (Theme – Rice is life)
2. 2005 : World Year of Physics/Year of Parthenium/Neem Patented Year (Patent From US)
3. 2006 : International Year of Desert and Desertification
4. 2007 : Water Year (theme – More Crop Per Drop), Year of Neeem
5. 2008 : International Potato Year and Sanitation (Food Safety and Quality Year (by Ministry of Food Processing Industries of India – 2008-09))
6. 2009 : International Astronomy Year
7. 2010 : International Year of Bio-Diversity; Year of Tigers (GOI)
8. 2011 : International Forest Year
9. 2012 : Horticulture Year (Declared on 17 Feb., 2012)
10. 2013 : U.N. International Year of Water Co-operation
11. 2013 : India Water Conservation Year (GOI) Theme – To Create Awareness About Water Conservation
12. 2013 : International Pulse Year (FAO/UNO)
13. 2014 : International Year of Family Farming (Declared by United Nations)

14. 2014 : Family Agriculture Year (NUO)
15. 2014 : Year of Farmer Producer Organization
16. 2016 : Pulse Year (IPTIC, Dubai)
17. 2017 : International Year of Sustainable Tourism for Development
18. 2018 : To be declared.

Descriptive (Epithet) Words for Fruit and Crops

- King of fruits - Mango
- Queen of fruits - Litchi
- King of spices - Pepper
- Queen of spices - Cardamom (Ilaychi)
- King of pulses - Gram
- Queen of pulses - Pea
- King of flowers - Rose
- Queen of flowers - Gladiolus
- King of temperate fruits - Apple
- King of arid fruit - Ber
- King of nuts - Almond
- King of vegetables - Potato
- A true fruit - Cashewnut
- Home of spices - India
- Poor man's fruit - Ber
- Miracle fruit - Kiwi
- Green gold - Afeem
- Queen of milk - Sinen goat
- Century plant - Dates plant (Khajur)
- Butter fruit - Avocado
- Nature's wonder - Chilli
- Black palm - Jamun
- Queen of cereals - Rice
- Bowl of Apple - J & K, HP
- Tree of heaven - Coconut
- Food of God - Cocoa
- Queen of beverage - Tea
- Adam's fig - Banana
- King of forest - Teak
- Man of the forest - Orangutan
- Camel crop - Jowar (Sorghum)

- The small wonder - Lemon
- Poor man's fruit - Guava (Amrud)
- Butter fruit - Avocado

Soil pH Requirement for Different Crops

pH 4.0-6.0	pH 5.0-6.5	pH 6.0-7.5	pH 5.0-7.5	pH 6.0-8.0
Potato	Apple	Apricot	Banana	Avocado
	Blackberry	Cherry	Rhubarb	Asparagus
	Cranberry	Grapevine	Strawberry	Ginger
	Gooseberry	Grapefruit	Raspberry	Leek
	Mango	Hazelnut	Carrot	Mint
	Melon	Hop	Cauliflower	Paprika
	Pineapple	Lemon	Sweet corn	Watercress
	Pomegranate	Lychee	Cucumber	
	Watermelon	Mulberry	Garlic	
	Basil	Nectarine	Lentil	
	Chicory	Peach	Parsley	
	Fennel	Plum	Pepper	
	Olive	Quince	Pumpkin	
	Peanut	Artichoke	Shallot	
	Sweet potato	Bean	Spearmint	
	Rice	Beetroot	Thyme	
	Rosemary	Broccoli	Tomato	
	Sage	Brussels sprouts	Turnip	
	Soyabean	Cabbage		
		Calabrese		
		Celery		
		Chinese cabbage		
		Chives		
		Lettuce		
		Millet		
		Mushroom		
		Mustard		
		Onion		
		Pea		
		Peppermint		
		Radish		
		Spinach		

Soil pH and Nutrient Availability

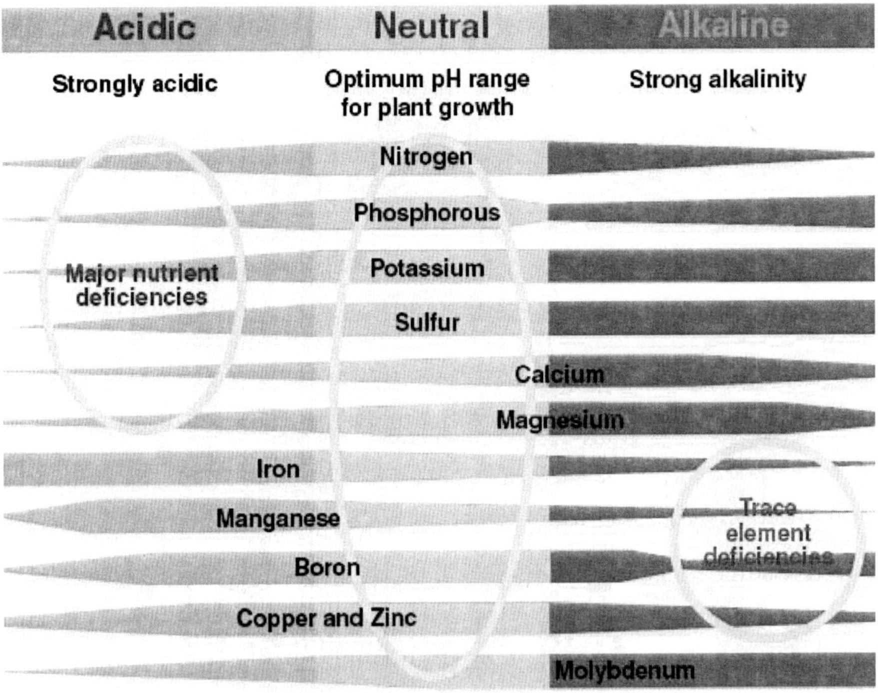

Agricultural Revolutions

1. **Green Revolution -** Food grains: Father of Green Revolution M.S. Swaminathan (1960s) HYV.
2. **White Revolution** (In India: Operation Flood) - Milk/Dairy production: 1970 Father of White Revolution Dr. Verghese Kurien 1st position, 1965, the National Dairy Development Board (NDDB).
3. **Blue Revolution** - Fish Production: 1960 Father of Blue Revolution Prof: Hiralal Chaudhuri.
4. **Yellow Revolution** - Oil Seeds Production: Father of Yellow Revolution Sam Pit Roda.
5. **Golden Revolution** - Fruits/Overall Horticulture Development/ Honey Production.
6. **Grey Revolution** - Fertilizer: 1964.
7. **Red Revolution** - Meat & Tomato Production
8. **Round Revolution** – Potato.

9. **Silver Fiber Revolution** – Cotton.
10. **Silver Revolution** - Egg/Poultry Production.
11. **Pink Revolution** - Onion Production/Pharmaceutical/Prawn Production.
12. **Brown Revolution** - Leather/Non-conventional/Cocoa Production.
13. **Golden Fiber Revolution** - Jute Production.
14. **Evergreen Revolution** - Overall Development of Agriculture.

Important Dates in Agriculture

1. 2 Feb – World wetland day
2. 8 March – International women's day
3. 11 March – Water resource day
4. 15 March — World consumer day
5. 21 March – World forest day
6. 22 March- World day for water
7. 23 March- World meteorological day
8. 7 April – World health day
9. 22 April – Earth day
10. 26 April - World intellectual property day
11. 1 May - World labour day
12. 11 May - National technology day
13. 31 May – World Anti-Tobacco day
14. 5 June - World environment day
15. 1 July – National agriculture day
16. 11 July - World population day
17. 16 July - ICAR day
18. 8 Sep - World literacy day
19. 16 Sep - World ozone day
20. 1 Oct - World habitat day
21. 2 Oct – National women literacy day / International Non-violence day

22. 4 Oct – World animal welfare day
23. 13 Oct – UN International day for natural disaster reduction
24. 16 Oct – World food day
25. 24 Oct - UN day
26. 16 Nov – National press day
27. 4 Dec – Agriculture women day
28. 10 Dec - Human rights day
29. 14 Dec – National energy conservation day
30. 23 Dec - Farmers day

Scientific Instruments Used in Agriculture

1. Lysimeter - water loss from soil
2. Rain gauge - from rainfall
3. Anemometer - speed of wind
4. Lactometer - density of milk
5. Aerometer - weight and density
6. Barometer - atmosphere pressure
7. Barograph - automatic measurement of change in pressure
8. Crescograph - recording growth of plants
9. Hydrometer - specific gravity of water
10. Hygrometer - relative humidity in atmosphere

Diseases Transmitted by Insects

White fly

Bhendi yellow mosaic

Chilly mosaic

Tomato spooted wilt

Tapioca mosaic

Cotton mosaic

Cotton leaf curl

Tobacco leaf curl

Aphid

Foorkey and katte disease of cardamom

Cucumber mosaic

Ground nut rosette virus

Citrus aphid

Citrus tristeza virus

Groundnut aphid

Papaya mosaic

Cowpea mosaic

Green peach aphid

Sugarbeet yellow mosaic

Potato virus X

Potato leaf roll

Banana aphid

Banana bunch top

Banana mosaic

Whorl aphid

Sugarcane mosaic

Thirps

Soyabean spotted wilt

Tomato spotted wilt

Chilly leaf curl

Groundnut bud necrosis

Eriophyid mite

Wheat streak mosaic

Red gram sterility mosaic

Sugarcane streak mosaic

Green leaf Hooper

Rice grassy stunt

Rice transitory yellowing

Zig zag leaf Hooper

Rice dwarf

Rice Orange leaf

Green leaf Hooper and Zig zag leaf Hooper

Rice tungro

Rice yellow Orange leaf

Brown plant Hooper

Rice ragged stunt

Leaf Hooper

Ragi mosaic

Sorghum shoot bug

Sorghum chlorosis

Corn mosaic

Freckled yellows of sorghum

Mealy bug

Sugar cane mottling

Pine apple root wilt

Psyllid

Citrus greening

Mango bud mite

Mango malformation

Mycoplasma diseases

Hopper

Sandle spike

Little leaf of brinjal

Leaf hooper

Sesamum phyllody

Area, Production and Yield of Major Crops

Crops	Area (Lakh hectare)			Production (Million Tonnes)			Yield (kg/hectare)		
	2014-15	2015-16	2016-17*	2014-15	2015-16	2016-17*	2014-15	2015-16	2016-17*
Rice	441.10	434.99	431.94	105.48	104.41	110.15	2391	2400	2550
Wheat	314.65	304.18	305.97	86.53	92.29	98.38	2750	3034	3216
Coarse cereals	251.70	243.89	247.71	42.86	38.52	44.19	1703	1579	1784
Pulses	235.54	249.12	294.65	17.15	16.35	22.95	728	656	779
Foodgrains	1243.00	1232.18	1280.26	252.02	251.57	275.68	2028	2042	2153
Oilseeds	255.96	260.87	262.06	27.51	25.25	32.10	1075	968	1225
Sugarcane	50.66	49.27	43.89	362.33	348.45	306.72	71512	70720	69886
Cotton@	128.19	122.92	108.45	34.80	30.01	33.09	462	415	519
Jute & Mesta#	8.10	7.82	7.66	11.13	10.52	10.60	2473	2421	2490

* 4th advance estimates @ Production in million bales of 170 kg each.

Production in million bales 180 Kg. each.

ICAR Awards, 2017

I. Chaudhary Devi Lal Outstanding All India Co-ordinated Research Project (AICRP) Award, 2017.

Awardees

Best All India Co-ordinated Research Project

All India Co-ordinated Research Project on Spices, Kozhikode, Kerala

Best Centre under this AICRP:

AICRP Centre-KAU Panniyur, Kerala

II. Rafi Ahmed Kidwai Award for Outstanding Research in Agricultural Sciences, 2017

Awardees

Crop & Horticultural Sciences Category

- Dr. D. K. Yadava, Head, Division of Seed Science & Technology, Indian Agricultural Research Institute, New Delhi & ADG, Seeds, ICAR, New Delhi Natural Resource Management & Agricultural Engineering Category
- Dr. B.S. Dwivedi, Head, Division of Soil Science & Agricultural Chemistry, ICAR-Indian Agricultural Research Institute, New Delhi.

Animal & Fisheries Sciences Category

- Dr. Kajal Chakraborty, Senior Scientist, Marine Bio-technology Division, ICAR-Central Marine Fisheries Research Institute. Ernakulum, Kochi, Kerala Social Sciences
- Dr. Anjani Kumar, International Scientist, International Food Policy Research Institute, South Asia Office, NASC Complex, New Delhi

III. Lal Bahadur Shastri Outstanding Young Scientist Award, 2017

Awardees

Crop & Horticultural Sciences

- Dr. B. Kalyana Babu, Scientist (Senior Scale), Agricultural Biotechnology, Indian Institute of Oil Palm Research, Pedavegi, Andhra Pradesh Natural Resource Management & Agricultural Engineering
- Dr. P.C. Abhilash, Assistant Professor (Stage II), Instt. of Environment & Sustainable Development, BHU, Varanasi Animal & Fisheries Sciences
- Dr. Rajendran Thomas, Sr. Scientist, ICAR-NRC on Pig, Rani, Guwahati, Assam Social Sciences
- Dr. Eldho Varghese, Scientist, Central Marine Fisheries Research Institute, Kochi, Kerala

IV. Jawaharlal Nehru Award for P.G. Outstanding Doctoral Thesis Research in Agricultural and Allied sciences, 2017

Awardees

Crop Sciences

1. R.K. Ellur, Scientist (Genetics & Plant Breeding), Division of Genetics, IARl, New Delhi
2. Dr. H.B. Mahesh, GKVK, Bengaluru (Karnataka) Animal Sciences:
1. Dr. Rajeev Ranjan, Scientist, International Centre for Foot & Mouth Disease, Jatni, Odisha
2. Dr. Sonika Ahlawat, National Bureau of Animal Genetic Resources, Karnal

Fisheries Science

1. Dr. Banya Kar, Patna (Bihar)
2. Dr. Anushree Velappan Nair, Marine Biology Division, CMFRl, Cochi, Kerala

Bio-technology

1. Dr. Asha Sriniwasan, Parippally, Kollam, Kerala
2. Dr. Amit Kumar Tripathi, Bhopal

Social Sciences

1. Dr. Anindita Datta, Paschim Vihar, New Delhi
2. Dr. Hema Baliwada, Scientist, Agricultural Extension, CTRI, Rajahmundry (A.P.)

Crop Protection

1. Dr. David Karamankodu Jacob, Scientist, Division of Insect Systematics, National Bureau of Agriculturally Important Micro-organisms, Bengaluru, Karnataka
2. Dr. Somnath Kadappa Holkar, Scientist & In-charge, IISR, Biological Control Centre, Pravarangar, Ahmadnagar.

Natural Resource Management

1. Dr. Bappa Das, Scientist, Agricultural Meteorology, ICAR-CCARI, Old Goa, North Goa
2. Dr. Saritha M., Scientist, Division of Integrated Land Use Management & Farming System, Central Arid Zone Research Institute, Jodhpur

Horticultural Sciences

1. Dr. Brij Bihari Sharma, Division of Vegetable Research, IARI, New Delhi
2. Dr. Selvakumar R., Scientist, Horticulture, CITH, Sirnagar (J&K)

Agricultural Engineering

1. Dr. Prashant Sampatrao Deshmukh, Senior Scientist, ICAR-CIRCOT, Mumbai
2. Dr. Kale Sakharam Jagan, Central Institute of Post-Harvest Engineering & Technology, Abohar Campus, Punjab

V. Bharat Ratna Dr. C. Subramaniam Award for Outstanding Teachers, 2017

Awardees

Crop & Horticultural Sciences

- Dr. Archana Sachdev, Principal Scientist, Division of Bio-chemistry, Indian Agricultural Research Institute, New Delhi
- Natural Resource Management & Agricultural Engineering:
- Dr. Tapan Jyoti Purakayastha, Principal Scientist, Division of Social Science & Agricultural Chemistry, Indian Agricultural Research Institute, New Delhi

Social Sciences

- Dr. Alka Singh. Professor & Principal Scientist, Division of Agricultural Economics, ICAR-Indian Agricultural Research Institute, New Delhi

VI. Panjabrao Deshmukh Outstanding Women Scientist Award, 2017

Awardees

I Award (Shared)

1. Dr. Shelly Praveen, Head & Principal Scientist, Division of Biochemistry, ICAR-Indian Agricultural Research Institute, New Delhi
2. Dr. Seema Jaggi, Principal Scientist and Professor (Agricultural Statistics), ICAR- Indian Agricultural Statistics Research Institute, New Delhi

II Award (also Shared)

1. Dr. Neeru Bhooshan, In-charge, Zonal Technology Management & Business Planning and Development Unit, Indian Agricultural Research Institute, New Delhi
2. Dr. P. S. Vimala Devi, Principal Scientist (Agricultural Entomology), ICAR-Indian Institute of Oilseeds Research, Rajendranagar, Hyderabad, Telengana

VII. NASI-ICAR Award for Innovation and Research on Farm Implements, 2017

Awardees

1. Dr. Akhilesh Kumar Singh, Principal Scientist (FMP), ICAR-Indian Institute of Sugarcane Research, Lucknow, U.P.
2. Dr. Sukhbir Singh, Senior Scientist (FMP), ICAR-Indian Institute of Sugarcane Research, Lucknow, U.P.

VIII. Vasant Rao Naik Award for Research Application in Agriculture, 2017

Team Leader

- Dr. M.S. Shisrahatti, Chief Scientist: (AICRP for Dryland Agriculture, Region-1, Agricultural Research Station, Vijayapura under University of Agricultural Sciences, Dharwad, Karnataka)

Associates

1. Mr. V.S. Surakod, Agronomist
2. Dr. S. T. Hundekar, Soil Scientist

3. Dr. Vijayakumar A. Giritamannavar, Plant Breeder
4. Dr. Vivek V. Angadi, Principal Scientist, Agronomy
5. Dr. S. B. Kalaghatagi, Professor & Head, College of Agriculture, Vijayapura
6. Dr. S.C. Alagundagi, Professor of Agronomy, UAS, Dharwad
7. Dr. S. B. Devaranavadgi, Chief Scientist

IX. Fakhruddin Ali Ahmed Award for Outstanding Research in Tribal Farming Systems, 2017

Awardees

Team I

1. Dr. Narendra Prakash (Team Leader), ICAR Research Complex for NEH Region, Manipur Centre, Imphal
2. Dr. Subhra Saikat Roy (Associate), ICAR Research Complex for NEH Region, Manipur Centre, Imphal
3. Dr. Meraj Alam Ansari (Associate), ICAR Research Complex for NEH Region, Manipur Centre, Imphal
4. Dr. Sushil Kumar Sharma (Associate), ICAR Research Complex for NEH Region, Manipur Centre, Imphal

Team II

1. Dr. Narendra Singh (Team Leader), Defence Research and Development Organization (DRDO), Defence Institute of High Altitude (DIHAR), Leh-Laddakh, (J&K)
2. Dr. Dorjay Angchuk (Associate), Defence Research and Development Organisation (DRDO), Defence Institute of High Altitude (DIHAR), Leh-Laddakh, (J&K)
3. Er. Sarfraj Nawaj (Associate), Defence Research and Development Organisation (DRDO), Defence Institute of High Altitude (DIHAR), Leh-Ladakh, (J&K)
4. Ms. Suman Tigga (Associate), Defence Research and Development Organisation (DRDO), Defence Institute of High Altitude (DIHAR), Leh-Laddakh, (J&K)

X. Swami Sahajanand Saraswati Outstanding Extension Scientist Award, 2017

Awardees

1. Dr. I. S. Tomar, Senior Scientist & Head, Rajmata Vijayaraje Scindia Krishi Vishwavidyalaya, Gwalior, Krishi Vigyan Kendra, Jhabua, M. P.
2. Shared equally between:
1. Dr. M. Kumaran, Principal Scientist (Agricultural, Extension), Social Science Division, ICAR-Central Institute of Brackish Water Aquaculture, Chennai, Tamil Nadu and;
2. Dr. D. R. Palsaniya, Senior Scientist, Agronomy, Crop Protection Division, Indian Grassland Fodder Research Institute" Jhansi. U.P.

XI. ICAR Cash Awards, 2017

Awardees

Administrative Category

1. Sh. Govind Prasad Sharma, Chief Finance and Accounts Officer. ICAR-Central Arid Zone Research Institute, Jodhpur, Rajasthan
2. Sh. Rajesh Kumar Jha, Under Secretary, Office of the Hon'ble Union Minister of Agriculture and Farmers Welfare, Krishi Bhawan, New Delhi
3. Sh. NVRN Murty, Finance and Accounts Officer, ICAR-ICAR-Central Institute of Freshwater Aquaculture, Bhubaneswar, Odisha

Technical Category

1. Dr. Ravi Kant, Assistant Chief Technical Officer, Animal Bio-chemistry Division, ICAR-National Dairy Research Institute, Karnal, Haryana
2. Sh. Devinder Singh, Technical Assistant (T-3), Division of Genetics, ICAR-Indian Agricultural Research Institute, New Delhi
3. Dr. Dhananjay Kumar Verma, Assistant Chief Technical Officer, Aquaculture Production and Environment Division, ICAR-Central Institute of Freshwater Aquaculture, Bhubaneshwar, Odisha
4. Sh. Om Parkash, Technical officer, ICAR-Indian Institute of Wheat and Barley Research, Karnal, Haryana

Supporting Category

1. Sh. B. Ashok, Skilled Support Staff National Academy of Agricultural Research Management, Hyderabad, Telangana

2. Sh. Aman Kumar, Skilled Support Staff, Indian of Wheat and Barley Research, Karnal, Haryana

XII. Award for Three Highest Ranked Universities

Awardees

- **Rank I:** National Dairy Research Institute, Deemed University, Karnal
- **Rank II:** Indian Agricultural Research Institute, Deemed University, New Delhi
- **Rank III:** Guru Angad Dev Veterinary and Animal Science University, Ludhiana.

❑❑❑

Chapter 15

Banking Awareness and AFO Interview Quick Guide

Basics of Banking Awareness and Banking Abbreviations

Q. What is Bank?

Bank is a financial institution which deals in debts and credits. It accepts deposits, lends money, etc. It bridges the gap between the savers and borrowers. **Primary functions of the banks include:**

1. Acceptance of deposits
2. Advancing loans
3. Creation of credit
4. Clearing of cheques
5. Financing foreign trade
6. Remittance of funds

Q. What do you mean by Central Bank of India?

The central bank of the country is the Reserve Bank of India (RBI). It was established in April, 1935 with a share capital of Rs. 5 crore based on the recommendations of the Hilton Young Commission. RBI was nationalized on 1 January, 1949.

The Management of RBI: The Executive head of the RBI is called the Governor, who is assisted by 4 Deputy Governors and other executive officers.

New Governor of RBI – Urjit Patel has been appointed as the new Governor of RBI. He assumed office as the 24th Governor of the Reserve Bank of India on September 5, 2016.

Note: Dr. Raghuram Rajan the 23rd Governor of the Reserve Bank of India, retired on September 4, 2016.

Deputy Governors of RBI - There are 4 Deputy Governors of RBI:

1. BP Kanungo
2. S S Mundra
3. N S Vishwanathan
4. Viral Acharya.

Q. What is Monetary Policy?

Monetary policy is how central banks manage the money supply to guide healthy economic growth. The money supply is credit, cash, checks, and money market mutual funds. The most important of these is credit, which includes loans, bonds, mortgages, and other agreements to repay.

Objectives of Monetary Policy

- The primary objective of central banks is to manage inflation. The second is to reduce unemployment once inflation has been controlled.
- Central bank reduces inflation by raising interest rates, selling securities through open market operations, and other measures to reduce liquidity.

Instrument of Monetary Policy

There are several direct and indirect instruments that are used in the implementation of monetary policy.

Open Market Operations (OMOs)

- It refers to buying and selling of government securities by RBI in the open market. It controls the money supply in the economy. When RBI sells govt. securities to banks, the lendable resources of the latter are reduced and banks are forced to reduce or contain their lending, thus curbing the money supply. When money supply is reduced, it results in increase in the interest rates, which tends to limit spending and investment.

- On the other hand, when RBI buys Govt. securities from banks, their lending resources are higher, which in turn encourage banks to lend more in the market and lending leads to increase in money supply. When money supply is increased, it results decline in the interest rates tends to promote spending and investment.

Cash Reserve Ratio (CRR)

It is the amount of funds that the banks have to keep with the RBI.

The Current CRR is 4%

For ex - When a bank's deposits increase by Rs. 100, and if the cash reserve ratio is 4%, the banks will have to hold additional Rs. 4 with RBI and Bank will be able to use only Rs. 96 for investments and lending/credit purpose. Therefore, higher the ratio (i.e. CRR), the lower is the amount that banks will be able to use for lending and investment.

Statutory Liquidity Ratio (SLR)

It indicates the minimum percentage of deposits that the bank has to maintain in the form of gold, cash or other approved securities.

The Current SLR is 19.50%

Note: If SLR increases, banks need to keep more liabilities (deposits) with them and provides less loans to people. If SLR decreases, banks need to keep fewer liabilities (deposits) with them and provides more loans to people. Changes in SLR often influence the availability of resources in the banking system for lending to the private sector.

Repo Rate

It is the rate at which RBI lends money to commercial banks in the event of any shortfall of funds. It is the rate of interest which RBI implements on the short-term loans, i.e., from a period ranging between 2 days to 3 months (90 Days). It is used by monetary authorities to control inflation. A reduction in the repo rate helps banks get money at a cheaper rate and vice versa.

The Current Repo Rate is 6.00%

Reverse Repo Rate

It is the rate at which the RBI borrows money from commercial banks. Banks are always happy to lend money to the RBI since their money is in safe hands with a good interest. An increase in reverse repo rate can prompt banks to park more funds with the RBI to earn higher returns on idle cash. It is also a tool which can be used by the RBI to drain excess money out of the banking system.

The Current Reverse Repo Rate is 5.75%

Marginal Standing Facility (MSF)

A facility under which scheduled commercial banks can borrow additional amount of overnight money from the Reserve Bank by dipping into their SLR portfolio upto a limit (currently two per cent of their net demand and time liabilities deposits) at a penal rate of interest (currently 100 basis points above the repo rate). This provides a safety valve against unanticipated liquidity shocks to the banking system. MSF rate and reverse repo rate determine the corridor for the daily movement in short-term money market interest rates.

The Current MSF is 6.25%

Bank Rate

Bank rate is the rate of interest implemented by RBI, when it lends money to a public sector bank on a long-term basis, i.e. from a period ranging from 90 days to 1 year. Under this definition, Bank Rate and Repo Rate seem to be similar terms because both are the interest rates at which RBI lends money to banks.

The Current Bank Rate is 6.25%

However, the Repo Rate is a short-term measure and refers to short-term loans used for controlling the amount of money in the market, whereas Bank Rate is a long-term measure and is governed by the long-term monetary policies of the RBI. Bank rate is also referred to as the discount rate and is the rate of interest, which central bank charges on the loans and advances to a commercial bank.

Current Policy & Reserve Rates

1. Repo Rate	6.00%
2. Reverse Repo	5.75%
3. CRR (Cash Reserve ratio)	4.00%
4. SLR (Statutory Liquidity Ratio)	19.50%
5. MSF (Marginal Standing Facility)	6.25%
6. Bank Rate	6.25%

Q. What is SWIFT?

SWIFT: - Society for worldwide Interbank financial telecommunication.

India was 74th Nation to join SWIFT Network.

SWIFT Code is a standard format of bank Identifier code. This code is used particularly in International transfer of money between banks. A majority of FOREX related message are sent to correspondent banks abroad through SWIFT. SWIFT Code consist **8 or 11** characters.

Q. When code is of 8 digits, it is referred to primary office

1st, 2nd, 3rd and 4th digit – bank code

5th and 6th digits – country code

7th and 8th digits - location code

9th, 10th and 11th digits - branch code (optional).

Q. Nationalization

Nationalization is the process of taking a private industry or private assets into public ownership by a national government or state.

Q. Need for Nationalization in India

- The banks mostly catered to the needs of large industries, big business houses.
- Sectors, such as agriculture, small scale industries and exports were lagging behind.
- The poor masses continued to be exploited by the moneylenders.
- Following this, in the year 1949, 1st January the Reserve Bank of India was nationalized.

14 Commercial banks were nationalized in 19th July, 1969

Smt. Indira Gandhi was the Prime Minister of India, in 1969. These were :

- Central Bank of India
- Bank of India
- Punjab National Bank
- Bank of Baroda
- United Commercial Bank
- Canara Bank
- Dena Bank
- United Bank
- Syndicate Bank
- Allahabad Bank
- Indian Bank
- Union Bank of India
- Bank of Maharashtra
- Indian Overseas Bank

Six more commercial banks were nationalized in April, 1980. These were:

- Andhra Bank
- Corporation Bank
- New Bank of India
- Oriental Bank of Commerce
- Punjab & Sindh Bank
- Vijaya Bank.

Meanwhile on the recommendation of **M. Narsimhan Committee, RRBs (Regional Rural Banks)** were formed on Oct 2, 1975. The objective behind the formation of RRBs was to serve large unserved population of rural areas and promoting financial inclusion.

Public Sector Banks

Public Sector Banks (PSBs) are banks where a majority stake (i.e. more than 50%) is held by a government. The shares of these banks are listed on stock exchanges. In 2011, IDBI bank was nationalized with a minimum capital of Rs. 500 cr. **For Ex** – Bank of Baroda, Punjab National Bank, Bank of India, etc.

Private Banks

The private sector banks in India represent part of the Indian banking sector that is made up of both private and public sector banks. The "private sector banks" are banks where greater parts of stake or equity are held by the private shareholders and not by the government. **For ex** – ICICI Bank, HDFC Bank, Axis Bank, etc.

Regional Rural Banks

Regional Rural Banks are local level banking organizations operating in different States of India. They have been created with a view to serve primarily the rural areas of India with basic banking and financial services. However, RRB's may have branches set up for urban operations and their area of operation may include urban areas too. The Government of India, the concerned State Government and the bank, which had sponsored the RRB contributed to the share capital of RRBs in the proportion of 50%, 15% and 35%, respectively.

Q. What Is KYC?

It is a process by which banks obtain information about the identity and address of the customers. This process helps to ensure that banks' services are not misused. The KYC procedure is to be completed by the banks while opening accounts and also periodically update the same.

KYC guidelines were introduced in the year 2002 by RBI and all banks were asked to make all accounts KYC complaiant by 31 December 2005. These guidelines are issued under Section 35 A of the Banking Regulation Act, 1949.

KYC Documents

These are the documents used to establish customer's identity. Banks need two types of document one for identity another for address

Q. What is Financial Inclusion?

Financial inclusion is a concept of making available banking/ financial services to a vast section of low income groups and weaker sections at an affordable price. The objective of financial inclusion is to provide the service of basic banking products to the unserved masses of the country, aiming towards inclusive economic growth.

The need for financial inclusion raised, when even after years of Independence 40% population lacked access to basic banking products. And apart from poverty, illiteracy and lack of regular income the major barriers are lack of reach, high cost of transactions and time.

Q. What is NABARD?

- NABARD was established on the recommendations of Shivaraman Committee on 12 July, 1982 to implement the National Bank for Agriculture and Rural Development Act, 1981. Its headquarter is located in Mumbai.
- It replaced the Agricultural Credit Department (ACD) and Rural Planning and Credit Cell (RPCC) of Reserve Bank of India, and Agricultural Refinance and Development Corporation (ARDC).
- It is one of the premiere agencies to provide credit in rural areas. NABARD is set up as an apex Development Bank with a mandate for facilitating credit flow for promotion and development of agriculture, small scale industries, cottage and village industries, handicrafts and other rural crafts.

Q. What is NPA?

An asset (loan), including a leased asset, becomes non-performing, when it stops generating income for the bank.

Note: Once the borrower has failed to make interest or principle payments for 90 days, the loan is considered to be a non-performing asset. It had been decided to adopt the '90 days' overdue' norm for identification of NPA, from the year ending March 31, 2004.

The conditions under which an asset becomes an NPA are as follows:

If interest or installment or both of principal remain overdue for a period of **more than 90 days** in respect of a term loan.

If Overdraft/ Cash Credit for an account remains **'out of order'**.

If bill remains overdue for a period of **more than 90 days** in the case of bills purchased and discounted. If installment of principal or interest **remains overdue for two crop seasons for short duration crops/one crop season for long duration crops.**

Asset Classification

RBI has classified non-performing assets into the following **three categories,** based on the period for which the asset has remained non-performing and the realizability of the dues:

Substandard Assets

With effect from March 31, 2005, a substandard asset is one, which has remained NPA for **a period less than or equal to 12 months.**

Doubtful Assets

With effect from March 31, 2005, an asset is classified as doubtful, if it has remained in the sub-standard category for a **period of 12 months**.

Loss Assets

A loss asset is one where loss has been identified by the bank or internal or external auditors or the RBI inspection but the amount has not been written off wholly. In other words, such an asset is considered **uncollectible** and of such little value that its continuance as a bankable asset is not warranted, although there may be some rescue or recovery value.

Negotiable Instrument Act, 1881

Negotiable instruments are written orders or unconditional promises to pay a fixed sum of money on demand or at a certain time. Negotiable instruments may be transferred from one person to another, who is known as a holder in due course. Upon transfer, also called negotiation of the instrument, the holder obtains full legal title to the instrument. Negotiable instruments may be transferred by delivery or by endorsement and delivery. For e.g. promissory notes, bills of exchange, cheques, drafts, certificates of deposit are all examples of negotiable instruments.

Cheque

Cheque is an instrument in writing containing an unconditional order, addressed to a banker, sign by the person who has deposited money with the banker, requiring him to pay on demand a certain sum of money only to or to the order of certain person or to the bearer of instrument.

Bearer Cheque

The bearer cheque is payable to the person specified therein or to any other else who presents it to the bank for payment. However, such cheques are risky, this is because if such cheques are lost, the finder of the cheque can collect payment from the bank.

Order Cheque

It is the one which is payable to a particular person. In such a cheque, the word 'bearer' may be cut out or cancelled and the word 'order' may be written. The payee can transfer an order cheque to someone else by signing his or her name on the back of it.

Crossed Cheque

When a cheque is crossed, the holder cannot encash it at the counter of the bank. The payment of such cheque is only credited to the bank account of the payee. Crossed cheque is done by drawing two parallel lines across top left corner of the cheque, with or without writing 'Account payee' in the space between the lines.

A Self Cheque

It is written by the account holder as pay self to receive money in physical form from the branch, where he holds his account. This can be alternated by using an ATM card.

Post-Dated Cheque (PDC)

A PDC is a form of a crossed or account payee bearer cheque but post-dated to meet the said financial payment at a future date. The cheque is valid from the date of issue to three months.

A Banker's Cheque

A banker's cheque is issued by a bank drawing money from its own funds rather than that from an account holder's. Banker's cheque is issued after the bank verifies the account status of the requestor and the amount is immediately deducted from the customer's account. A banker's cheque cannot be dishonoured as in the case of a normal cheque, when an account holder has insufficient funds in his/her account. Though different from a normal cheque, it requires clearing too.

Stale Cheque - If a cheque is presented for payment after three months from the date of the cheque, it is called stale cheque. A stale cheque is not honoured by the bank.

Q. What is RTGS System?

RTGS' stands for Real Time Gross Settlement, which can be defined as the continuous (real-time) settlement of funds transfers individually on an order by order basis (without netting). 'Real Time' means the processing of instructions at the time they are received rather than at some later time. 'Gross Settlement' means the settlement of funds transfer instructions occurs individually (on an instruction by instruction basis). Considering that the funds settlement takes place in the books of the Reserve Bank of India, the payments are final and irrevocable.

Q. What is NEFT?

National Electronic Funds Transfer (NEFT) is a nation-wide payment system facilitating one-to-one funds transfer. Under this Scheme, individuals can electronically transfer funds from any bank branch to any individual having an account with any other bank branch in the country participating in the Scheme.

RTGS Vs NEFT

Thus, we can say that both RTGS and NEFT are schemes started by RBI for the benefit of the customers which allow accounts holders in the banks to electronically transfer the funds intra-bank. In the case of RTGS, settlement in on 'Real Time' basis, whereas in case of NEFT, the settlement in on batch basis and net basis. There are some other rules, regulations and differences, which we will be discussing below.

Q. Different Types of Bank Accounts

Normally there are 4 types of deposits account in Bank which are:

- Savings Account
- Current Account
- Recurring Account
- Fixed Deposit Account

Difference b/w Saving & Current Account– Savings Bank Account	**Current Bank Account**
Bank pays interest on daily basis	Bank does not pay any interest on current account.
Restriction on the number and amount of withdrawals	There is no restriction on the number and amount of withdrawals
SA can be opened by individuals, small businesses and students, etc.	Current Accounts are for big businesses, companies and institutions, etc.
Nomination facility is available for SA deposits.	Nomination facility is normally not available for CA deposits.

Note: Now **CA & SA together are known as demand deposit,** since they are payable on demand. Suppose you need Rs.5000 today you withdraw Rs.5000 today.

CASA RATIO – The Ratio of the deposits in the form of **Current Account & Saving Account to the total deposits is known as CASA Ratio** (normally expressed in %). More CASA ratio banks are in the safe zone. ICCI, AXIS, HDFC & SBI have CASA Ratio more than 40%.

Fixed Deposit and Recurring Deposit - these deposits are **TIME** deposits (as for specified agreed period b/w you and the bank). In these deposits, interest paid by the bank is slightly higher than CA & SA **Note**: If you are depositing the money **for 1 year, you will get the money after 1 year**. But if you withdraw before the agreed period i.e. before 1 year, you have to pay penalty. **Cheque book facility is not available for time deposits**.

Banking Abbrevations

1. **PSBs**: Public Sector Banks
2. **SNBCs**: Schedule Non-Commercial Banks

3. **SENSEX**: Sensitive Index of Stock Exchange
4. **GNP**: Gross National Product
5. **KYC**: Know Your Customer
6. **RTGS**: Real Time Gross Settlement
7. **NEFT**: National Electronic Money Transfer
8. **EFT**: Electronic Fund Transfer
9. **CBS**: Core Banking Solutions
10. **LIBOR**: London Interbank Offered Rate
11. **MIBOR**: Mumbai Interbank Offered Rate
12. **MIBID**: Mumbai Interbank Bid Rate
13. **SARFAESI**: Securitisation and Reconstruction of Financial Assets and Enforcement of Security Interest
14. **CAR**: Capital Adequecy Ratio
15. **FIIs**: Foreign Institutional Investments
16. **FDI**: Foreign Direct Investment
17. **IPO**: Initial Public Offering
18. **MICR**: Magnetic Ink Character Recognition
19. **BIRD**: Bankers Institute of Rural Development
20. **IBA**: Indian Bank Association
21. **BPLR**: Benchmark Prime Lending Rate
22. **ICICI**: Industrial Credit and Investment Corporation of India
23. **HDFC**: Housing Development Finance Corporation
24. **SWOT**: Strength, Weaknesses, Opportunities and Threats
25. **SWIFT**: Society for Worldwide Interbank Financial Telecommunication
26. **FERA**: Foreign Exchange Regulatory Act
27. **FEMA**: Foreign Exchange Management Act
28. **CASA**: Current and Saving Account
29. **NDTL**: Net Demand and Time Liabilities
30. **NASDAQ**: National Association for Securities Dealers Automated Quotations
31. **CRISIL**: Credit Rating and Investment Services India Limited

32. **CIBIL**: Credit Information Bureau of India Limited
33. **NAV**: Net Asset Value
34. **ICRA**: Indian Credit Rating Agency
35. **CARE**: Credit Analysis and Research Limited
36. **WMAs**: Ways and Means Advances
37. **ALM**: Asset Liability Management
38. **INFINET:** Indian Financial Network
39. **OLTAS** - On-line Tax Accounting System (OLTAS) for Direct Taxes
40. **TIN** - Tax Information Network (TIN)
41. **IMPS** - Interbank Mobile Payment Service (IMPS) or Immediate Payment Service
42. **CDR**- Corporate Debt Restructuring
43. **CAD**- Capital Account Deficit
44. **REITs**: Real Estate Investment Trusts
45. **InvITs**: Infrastructure Investment Trusts
46. **ALM**- Asset Liability Management
47. **ASBA**: Application Supported by Blocked Amount
48. **PIN**: Personal Identification Number
49. **CCEA** – Cabinet Committee on Economic Affairs
50. **CECA** - Comprehensive Economic Co-operation Agreement
51. **CEPA** – Comprehensive Economic Partnership Agreement
52. **DTAA** – Double Taxation Avoidance Agreement
53. **ECBs** - External Commercial Borrowings
54. **EFSF** – European Financial Stability Facility
55. **FINO**- Financial Inclusion Network Operation
56. **FIPB** – Foreign Investment Promotion Board
57. **FSLRC** – Financial Sector Legislative Reforms Commission
58. **CRAR**: Capital to Risk-weighted Assets Ratio
59. **LCR**: Liquidity Coverage Ratio
60. **TARC** - Tax Administration Reform Commission

61. **GIRO** - Government Internal Revenue Order
62. **FRBMA**: Fiscal Responsibility and Budget Management Act
63. **AMFI-** Association of Mutual Fund in India.
64. **TIEA** – Tax Information eExchange Agreement
65. **GAAR** - General Anti-Avoidance Rule
66. **GSLV** - Geo-Synchronous Launch Vehicle
67. **PPP** – Public Private Partnership & Purchasing Power Parity
68. **PSLV** – Polar Satellite Launch Vehicle
69. **TAPI** - Turkmenistan-Afghanistan-Pakistan-India.
70. **QFI** -Qualified Foreign Investors
71. **AD**-Authorized Dealer.
72. **ASSOCHAM**-Associated Chambers of Commerce and Industry of India.
73. **BCSBI**-Banking Codes and Standard Board of India.
74. **BIS**-Bank for International Settlements.
75. **CDS**-Credit Default Swap.
76. **CEPA-**Comprehensive Economic Partnership Management.
77. **FIMMDA**-Fixed Income Money Markets and Derivatives Association.
78. **FPI**-Foreign Portfolio Investment.
79. **IBRD**-International Bank for Reconstruction And Development.
80. **UIDAI**-Unique Identification Development Authority of India

Important Points to take care of before entering your interview panel

Do's

- Ask for permission before entering: Sir, May I come in?
- Say thank you after their permission.
- Softly close the door and wish the Panel members with gentle standing posture.
- Remember time carefully and wish accordingly, i.e. Good morning, good afternoon or good evening.

- Do remember Before 12 o'clock- Good Morning/Between 12-4 Good afternoon/After 4 pm- Good evening
- If there is any female member in the Panel, then greet her first.
- Wait till Panel asks/permits you to sit and say thanks before sitting.
- Sit in a straight posture with hands in your lap/thigh. Avoid fidgeting and slouching. Make your eye contact with every panel member.
- Make sure that your voice is very Clear/Audible/Humble and Polite. Give answers confidently.
- Be positive, honest and be yourself. Don't try to show off. Always say thank you sir, when Panel informs/ corrects you anywhere.
- When you are unable to answer any question say, Sorry sir, I do not know.
- When board say thanks, your interview is over- Then, say Thank you sir.

Don'ts

- Don't make excuses. Take responsibility for your decisions and your actions.
- Do not make negative comments about your previous employer.
- Do not treat the interview casually.
- Do not give impression that you are interested in particular location only and do not talk about salary and perks until the interviewer asks/enquire for.
- Don't assume that female interviewer is Mrs. or Miss. Address her as Ma'am unless told otherwise.
- Don't allow your cellphone to ring during interview. Put it Switch off or on Silent mode.
- Avoid hand movement.
- Avoid looking up and down and avoid hiding eyes with the panel members, as it shows lack of confidence.
- Don't forget to wish the Interview Panel members.
- Don't sit until the Interviewers ask to do so.